Lecture Notes in Computer Science 13026

T0171859

More information about this subseries at http://www.springer.com/series/7407

Hongfei Lin · Min Zhang ·
Liang Pang (Eds.)

Information Retrieval

27th China Conference, CCIR 2021
Dalian, China, October 29–31, 2021
Proceedings

Springer

Editors
Hongfei Lin ⓘ
Dalian University of Technology
Dalian, China

Liang Pang
Institute of Computing Technology
Chinese Academy of Sciences
Beijing, China

Min Zhang
Department of Computer Science
Tsinghua University
Beijing, China

ISSN 0302-9743   ISSN 1611-3349 (electronic)
Lecture Notes in Computer Science
ISBN 978-3-030-88188-7   ISBN 978-3-030-88189-4 (eBook)
https://doi.org/10.1007/978-3-030-88189-4

LNCS Sublibrary: SL1 – Theoretical Computer Science and General Issues

This Springer imprint is published by the registered company Springer Nature Switzerland AG
The registered company address is: Gewerbestrasse 11, 6330 Cham, Switzerland

Preface

The 2021 China Conference on Information Retrieval (CCIR 2021), co-organized by the Chinese Information Processing Society of China (CIPS) and the Chinese Computer Federation (CCF), was the 27th installment of the conference series. The conference was hosted by Dalian University of Foreign Languages in Dalian, Liaoning, China, during October 29–31, 2021.

The annual CCIR conference serves as the major forum for researchers and practitioners from both China and other Asian countries/regions to share their ideas, present new research results, and demonstrate new systems and techniques in the broad field of information retrieval (IR). Since CCIR 2017, the conference has enjoyed contributions spanning the theory and application of IR, both in English and Chinese.

This year we received a total of 124 submissions from both China and other Asian countries. Each submission was carefully reviewed by at least three domain experts, and the Program Committee (PC) chairs made the final decisions. We accepted 72, among which 15 were English papers and 57 were Chinese papers. The final English program of CCIR 2021 featured 15 papers.

CCIR 2021 included abundant academic activities. Besides keynote speeches delivered by world-renowned scientists from China and abroad, and traditional paper presentation sessions and poster sessions, we also hosted a young scientist forum, an evaluation workshop, and tutorials on frontier research topics. We also invited authors from related international conferences (such as SIGIR and CIKM) to share their research results as well. CCIR 2021 featured four keynote speeches by James Allen (University of Massachusetts Amherst), Si Wu (Peking University), Jun Wang (UCL), and Zhongyuan Wang (Meituan Inc.).

The conference and program chairs of CCIR 2021 extend their sincere gratitude to all authors and contributors to this year's conference. We are also grateful to the PC members for their reviewing effort, which guaranteed that CCIR 2021 could feature a quality program of original and innovative research in IR. Special thanks go to our sponsors for their generosity: Meituan Inc., Huawei Inc., and Baidu Inc.

October 2021

Hong Liu
Jiafeng Guo
Hongfei Lin
Min Zhang
Liang Pang

Organization

General Chairs

Hong Liu   Dalian University of Foreign Languages, China
Jiafeng Guo   Institute of Computing Technology, Chinese Academy of Sciences, China

Program Committee Chairs

Hongfei Lin   Dalian University of Technology, China
Min Zhang   Tsinghua University, China

Proceedings Chairs

Ruihua Qi   Dalian University of Foreign Languages, China
Liang Yang   Dalian University of Technology, China

Publicity Chair

Zhumin Chen   Shandong University, China

Publication Chair

Liang Pang   Institute of Computing Technology, Chinese Academy of Sciences, China

Webmaster

Yuan Lin   Dalian University of Technology, China

Youth Forum Chairs

Xin Zhao   Renmin University of China, China
Chenliang Li   Wuhan University, China

CCIR Cup Chairs

Jianming Lv   South China University of Technology, China
Weiran Xu   Beijing University of Posts and Telecommunications, China

Sponsorship Chairs

Qi Zhang   Fudan University, China
Zhongyuan Wang   Kwai Inc., China

Treasurers

Kan Xu   Dalian University of Technology, China
Song Yang   Dalian University of Foreign Languages, China

Award Chair

Ru Li   Shanxi University, China

Program Committee

Ting Bai   Beijing University of Posts and Telecommunications,
                        China
Fei Cai   National University of Defense Technology, China
Jiawei Chen   University of Science and Technology of China, China
Xiaoliang Chen   Xihua University, China
Yubo Chen   Institute of Automation, Chinese Academy of Sciences,
                        China
Zhumin Chen   Shandong University, China
Zhicheng Dou   Renmin University of China, China
Yajun Du   Xihua University, China
Yixing Fan   Institute of Computing Technology, Chinese Academy
                        of Sciences, China
Shengxiang Gao   Kunming University of Science and Technology, China
Zhongyuan Han   Foshan University, China
Yanbin Hao   University of Science and Technology of China, China
Ben He   University of Chinese Academy of Sciences, China
Xiangnan He   University of Science and Technology of China, China
Yu Hong   Soochow University, China
Zhuoren Jiang   Zhejiang University, China
Ting Jin   Hainan University, China
Yanyan Lan   Tsinghua University, China
Chenliang Li   Wuhan University, China
Lishuang Li   Dalian University of Technology, China
Ru Li   Shanxi University, China
Shangsong Liang   Sun Yat-sen University, China
Xiangwen Liao   Fuzhou University, China
Hongfei Lin   Dalian University of Technology, China
Yuan Lin   Dalian University of Technology, China
Chang Liu   Peking University, China
Peiyu Liu   Shandong Normal University, China

Yue Liu	Institute of Computing Technology, Chinese Academy of Sciences, China
Cheng Luo	Tsinghua University, China
Zhunchen Luo	PLA Academy of Military Science, China
Jianming Lv	South China University of Technology, China
Ziyu Lyu	Shenzhen Institute of Advanced Technology, Chinese Academy of Sciences, China
Weizhi Ma	Tsinghua University, China
Jiaxin Mao	Tsinghua University, China
Xianling Mao	Beijing Institute of Technology, China
Liqiang Nie	Shandong University, China
Liang Pang	Institute of Computing Technology, Chinese Academy of Sciences, China
Zhaochun Ren	Shandong University, China
Tong Ruan	East China University of Science and Technology, China
Huawei Shen	Institute of Computing Technology, Chinese Academy of Sciences, China
Dawei Song	Beijing Institute of Technology, China
Ruihua Song	Renmin University of China, China
Xuemeng Song	Shandong University, China
Hongye Tan	Shanxi University, China
Songbo Tan	Beijing Hengchang Litong Investment Management Co., Ltd., China
Liang Tang	Information Engineering University, China
Hongjun Wang	TRS Information Technology Co., Ltd., China
Pengfei Wang	Beijing Institute of Technology, China
Suge Wang	Shanxi University, China
Ting Wang	National University of Defense Technology, China
Zhiqiang Wang	Shanxi University, China
Zhongyuan Wang	Meituan, China
Dan Wu	Wuhan University, China
Le Wu	Hefei University of Technology, China
Yueyue Wu	Tsinghua University, China
Jun Xu	Renmin University of China, China
Tong Xu	University of Science and Technology of China, China
Weiran Xu	Beijing Institute of Technology, China
Hongbo Xu	Institute of Computing Technology, Chinese Academy of Sciences, China
Kan Xu	Dalian University of Technology, China
Hongfei Yan	Peking University, China
Xiaohui Yan	Huawei, China
Muyun Yang	Harbin Institute of Technology, China
Zhihao Yang	Dalian University of Technology, China
Dongyu Zhang	Dalian University of Technology, China
Hu Zhang	Shanxi University, China

Min Zhang	Tsinghua University, China
Peng Zhang	Tianjin University, China
Qi Zhang	Fudan University, China
Ruqing Zhang	Institute of Computing Technology, Chinese Academy of Sciences, China
Weinan Zhang	Harbin Institute of Technology, China
Ying Zhang	Nankai University, China
Yu Zhang	Harbin Institute of Technology, China
Chengzhi Zhang	Nanjing University of Science and Technology, China
Xin Zhao	Renmin University of China, China
Jianxing Zheng	Shanxi University, China
Qingqing Zhou	Nanjing Normal University, China
Jianke Zhu	Zhejiang University, China
Xiaofei Zhu	Chongqing University of Technology, China
Zhenfang Zhu	ShanDong JiaoTong University, China
Jiali Zuo	Jiangxi Normal University, China

Contents

Search and Recommendation

Interaction-Based Document Matching for Implicit Search Result
Diversification 3
   Xubo Qin, Zhicheng Dou, Yutao Zhu, and Ji-Rong Wen

Various Legal Factors Extraction Based on Machine Reading
Comprehension 16
   *Beichen Wang, Ziyue Wang, Baoxin Wang, Dayong Wu, Zhigang Chen,
Shijin Wang, and Guoping Hu*

Meta-learned ID Embeddings for Online Inductive Recommendation 32
   Jingyu Peng, Le Wu, Peijie Sun, and Meng Wang

Modelling Dynamic Item Complementarity with Graph Neural Network
for Recommendation 45
   Yingwai Shiu, Weizhi Ma, Min Zhang, Yiqun Liu, and Shaoping Ma

NLP for IR

LDA-Transformer Model in Chinese Poetry Authorship Attribution 59
   Zhou Ai, Zhang Yijia, Wei Hao, and Lu Mingyu

Aspect Fusion Graph Convolutional Networks for Aspect-Based Sentiment
Analysis 74
   Fuyao Zhang, Yijia Zhang, Shuo Hou, Fei Chen, and Mingyu Lu

Iterative Strict Density-Based Clustering for News Stream 88
   Kaijie Shi, Jiaxin Shi, Yu Zhou, Lei Hou, and Juanzi Li

A Pre-LN Transformer Network Model with Lexical Features
for Fine-Grained Sentiment Classification 100
   Kaixin Wang, Xiujuan Xu, Yu Liu, and Zhehuan Zhao

Adversarial Context-Aware Representation Learning
of Multiword Expressions 112
   Bo An

IR in Education

Research on the Evaluation Words Recognition in Scholarly Papers'
Peer Review Texts 129
   Kun Ding, Xinhang Zhao, Liang Yang, Kaiqiao Wang, and Yuan Lin

Evaluation of Learning Effect Based on Online Data. 141
   Zhaohui Liu, Hongfei Yan, Chong Chen, and Qi Su

Self-training vs Pre-trained Embeddings for Automatic Essay Scoring 155
   *Xianbing Zhou, Liang Yang, Xiaochao Fan, Ge Ren, Yong Yang,
   and Hongfei Lin*

Enhanced Hierarchical Structure Features for Automated Essay Scoring. 168
   Junteng Ma, Xia Li, Minping Chen, and Weigeng Yang

IR in Biomedicine

A Drug Repositioning Method Based on Heterogeneous Graph Neural
Network. 183
   Yu Wang, Shaowu Zhang, Yijia Zhang, Liang Yang, and Hongfei Lin

Auto-learning Convolution-Based Graph Convolutional Network
for Medical Relation Extraction 195
   *Mengyuan Qian, Jian Wang, Hongfei Lin, Di Zhao, Yijia Zhang,
   Wentai Tang, and Zhihao Yang*

Author Index 209

Search and Recommendation

Interaction-Based Document Matching for Implicit Search Result Diversification

Xubo Qin[1], Zhicheng Dou[2(✉)], Yutao Zhu[3], and Ji-Rong Wen[2]

[1] School of Information, Renmin University of China, Beijing, China
[2] Gaoling School of Artificial Intelligence, Renmin University of China,
Beijing, China
dou@ruc.edu.cn
[3] Université de Montréal, Québec, Canada

Abstract. To satisfy different intents behind the queries issued by users, the search engines need to re-rank the search result documents for diversification. Most of previous approaches of search result diversification use pre-trained embeddings to represent the candidate documents. These representation-based approaches lose fine-grained matching signals. In this paper, we propose a new supervised framework leveraging interaction-based neural matching signals for implicit search result diversification. Compared with previous works, our proposed framework can capture and aggregate fine-grained matching signals between each candidate document and selected document sequences, and improve the performance of implicit search result diversification. Experimental results show that our proposed framework can outperform previous state-of-the-art implicit and explicit diversification approaches significantly, and even slightly outperforms ensemble diversification approaches. Besides, with our proposed strategies the online ranking latency of our framework is moderate and affordable.

Keywords: Search result diversification · Neural IR · Matching

1 Introduction

Users tend to issue short queries in search engines. These short queries are usually ambiguous or vague [12,16,26,27]. Taking the query "apple" as an example, the actual user intents behind the query can be either the fruit "apple" or "Apple Company". Besides, a user intent can also cover multiple aspects (such as "how to learn JAVA" or "download JAVA IDE" for the intent "JAVA programming language"). In order to satisfy those diversified user intents, the technology of search result diversification is necessary for search engines. The ranking models of search result diversification aims at re-ranking the result documents to satisfy diversified user intents at former ranking positions. Depending on whether to model the user intent coverage explicitly, previous studies can be categorized into implicit and explicit diversification methods. The implicit diverse ranking

© Springer Nature Switzerland AG 2021
H. Lin et al. (Eds.): CCIR 2021, LNCS 13026, pp. 3–15, 2021.
https://doi.org/10.1007/978-3-030-88189-4_1

approaches [4,30,31,33,36] focus on capturing the interaction signals between documents and modeling the document novelty by the dissimilarity of the documents. On the contrary, the explicit approaches [1,9,14,17,24] tend to explicitly model the coverage of different subtopics. Recently, a group of studies [21,22] are proposed for modeling both document interactions and subtopic coverage, which can be treated as ensemble methods. As subtopic mining itself is a very challenging task, in this work, we focus on implicit diversification approaches.

Although many implicit methods have been proposed, most of them measure the document's novelty based on the dissimilarity between the candidate document and the selected documents. For example, NTN [31] is a typical implicit method that automatically learns a novelty function based on the pre-trained representation (*e.g.*, doc2vec or PLSA) of documents. A main drawback of these methods is: merely computing the document's novelty based on the pre-trained representation is inaccurate, because the unsupervised pre-training methods cannot provide reliable representations, and the document's content usually contains abundant information. Indeed, some studies in ad-hoc ranking [13] have reported that the representation-based methods (*i.e.*, directly computing ranking score based on the representation of queries and documents) often performs worse than interaction-based methods (*i.e.*, constructing the term-level matching signals from queries and documents and aggregating them for calculating ranking scores). This result indicates that merely using pre-trained document representation to compute documents' similarity is suboptimal in search result diversification.

To tackle this problem, in this work, we propose conducting the **term-level interaction between documents** to measure their similarity and design a new model called **MatchingDIV**. Our model follows the widely used greedy document selection process in search result diversification. Given a document list, the model iteratively selects a novel document from the list and adds it to the re-ranked list. After all documents are selected, the obtained list is diverse. Specifically, MatchingDIV first encodes the candidate document and all selected documents by a pre-trained language model (*e.g.*, BERT [11]). Then, each selected document is interacted with the candidate document at the term-level, and the representation of each term in the selected document is updated. By this means, the fine-grained matching information is integrated into the term representations. Next, MatchingDIV applies a recurrent neural network (RNN) to aggregate the term-level representation and calculate the document-level representation. Finally, all document representations are aggregated by another RNN, and the ranking score of the candidate document is computed based on the final representation. To our best knowledge, we are the first to consider fine-grained interaction between documents in search result diversification task. Experiment results show that our proposed framework can significantly outperforms the state-of-the-art implicit and explicit diversification approaches based on pre-trained document representations.

2   Related Work

We briefly review some related work about search result diversification and neural matching in different tasks.

Search Result Diversification. The earliest typical diversification models is Max Margin Relevance (MMR) [4]. It compares each candidate document with selected document sequence, greedily selects the document with the best ranking score, and appends it to the selected sequence. The ranking scores of documents are computed based on their relevance to the query and the novelty compared with the selected documents. The "novelty" here is measured by the dissimilarity between documents. The original MMR uses handcrafted features and scoring functions to calculate the similarity, which limits its application. Many approaches extend MMR by using supervised learning methods to learn the features and functions automatically (*e.g.*, SVM-DIV [33], R-LTR [36], PAMM [30], and PAMM-NTN [31]). These methods are called *implicit diversification approaches*. On the contrary, *explicit diversification approaches* model the coverage of different user intents (represented as subtopics) of each document. A novel document is expected to cover new user intents which have not been covered by selected sequence. Several unsupervised and supervised explicit diversification approaches have been proposed, *e.g.*, xQuAD [24], PM2 [9], HxQuAD, HPM2 [14], and DSSA [17]. Recently, a group of new approaches (*e.g.*, DVGAN [21] and DESA [22]) have been proposed as *ensemble approaches*. They use both implicit inter-document features and explicit subtopic coverage features.

Note that most of these approaches use document representations pre-trained by unsupervised tools, such as doc2vec [20] and LDA [3]. The document similarity is computed by cosine similarity of two document embeddings. Different from these methods, we represent documents at term-level, based on which an interaction is conducted. Therefore, our method can capture fine-grained matching signals which are more accurate in computing document similarity.

Neural Matching in Different Tasks. In recent years, researchers have proposed a group of deep-learning based relevance matching models for multiple IR tasks. Compared with traditional approaches, these neural matching methods can better measure the semantic similarity between queries and documents. In general, these methods can be divided into two categories: representation-based methods [15,25] and interaction-based methods [8,32]. The representation-based methods use neural networks to generate the dense vector representations of the queries and documents and compute their similarity based on the representations. In contrast, the interaction-based methods first capture term-level interaction signals between queries and documents and then aggregate them to compute the similarity. In the view of neural matching, the previous diversification approaches can be seen as representation-based approaches. In addition to the ranking task, there are also a group of studies [28,35] leveraging neural models to measure the similarity between dialogue context and response candidate and achieving great performance in retrieval-based chatbots. Intuitively, the relationship among the context-response sentences is similar to the that among selected-candidate documents in implicit search result diversification tasks. Inspired by previous work in multi-turn response selection, we propose an interaction-based methods for search result diversification.

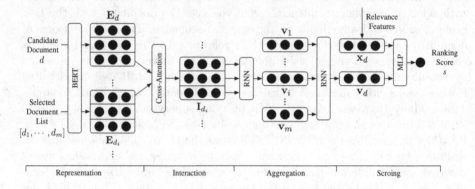

Fig. 1. The structure of MatchingDIV.

3 Methodology

In this section, we first give the formulation of search result diversification problem. Then, we introduce the overall structure of our framework and describe the details of each component. Finally, we describe the training and inference process.

3.1 Problem Formulation

The definition of implicit search result diversification task can be described as: Given a query q with and a list of candidate documents $\mathcal{D}$, the diverse ranking task aims to return a new ranked document list $\mathcal{R}$. Here, $\mathcal{D}$ is an initial relevance ranking list without diversification. For the diversified list $\mathcal{R}$, both the relevance and the diversity of those documents should be considered. As a greedy selection approach, our framework compares each candidate document d with the selected document sequence $\mathcal{C}$ and returns the ranking score s. The document with the highest score will be selected and appended into $\mathcal{C}$. Our target is designing a model f to compute the ranking score s for the candidate document d by considering its relevance to the query q and its novelty regarding $\mathcal{C}$. This process can be formulated as:

$$s = f(q, d, \mathcal{C}). \tag{1}$$

3.2   MatchingDIV

In implicit search result diversification methods, a document's novelty is measured by its dissimilarity with other documents. Therefore, how to calculate the similarity is very crucial. Existing methods usually compute cosine similarity based on pre-trained document representation, but it is difficult to capture accurate matching signals merely based on these representations. In this work, we propose an interaction-based document matching framework, which is called MatchingDIV. As shown in Fig. 1, our framework first represents each term of document by a pre-trained language model. Then, we design a cross-attention mechanism to model the interaction based on two documents' representations. Since this operation is conducted on term-level representations, our method can capture fine-grained matching signals. The details of our framework are introduced as follows.

Document Representation. With the recent progress of contextualized language models, we use BERT [11] to generate the term representation of the documents:

$$\mathbf{E}_d = \text{Linear}(\text{Norm}(\text{BERT}([D]))), \tag{2}$$

where $\mathbf{E}_d \in \mathbb{R}^{l_d \times h}$, and l_d is the length of the document. $[D]$ denotes the word-pieces of documents after tokenization, and $\text{Norm}(\cdot)$ denotes the operation of normalization. "BERT" denotes a BERT-like encoder which can be replaced by other pretrained models, such as DistilBERT [23] or ELECTRA [6]. Following the previous work [18], we apply a linear projection layer to compress the term representation into h-dimension for reducing the storage cost.

Interaction via Cross-Attention. To capture the fine-grained matching signals, we use cross-attention as the interaction function to let each document in the selected document sequence interact with the candidate document. Similar to the self-attention which is widely used in Transformer-based models [11,29], the cross-attention operation is also based on multi-head attention (MHA):

$$\text{Attn}(\mathbf{q}, \mathbf{K}, \mathbf{V}) = \text{Softmax}(\frac{\mathbf{q}\mathbf{K}^\top}{\sqrt{d}})\mathbf{V}, \tag{3}$$

$$\text{MHA}(\mathbf{q}, \mathbf{K}, \mathbf{V}) = [\mathbf{a}_1; \dots; \mathbf{a}_h], \tag{4}$$

$$\mathbf{a}_i = \text{Attn}(\mathbf{q}\mathbf{W}_i^Q, \mathbf{K}\mathbf{W}_i^K, \mathbf{V}\mathbf{W}_i^V), \quad i \in [1, h]. \tag{5}$$

Due to space limitation, we omit the details of multi-head attention, which can be referred to at [29]. For the selected document d_i in the sequence $\mathcal{C}$ and the candidate document d, the interacted representations of d_i can be defined as:

$$\mathbf{I}_{d_i} = \text{MHA}(\mathbf{E}_{d_i}, \mathbf{E}_d, \mathbf{E}_d). \tag{6}$$

With the cross-attention mechanism, the representation of each term in the selected document d_i is enhanced by the weighed sum of the representations of d.

The similarity information is updated to the representation so that the interacted representations $\mathbf{I}_{d_i}$ can represent the term-level matching signals between d and d_i.

Matching Signals Aggregation. After getting the enhanced representation of each term in the selected document d_i, the next question is how to aggregate them and compute an integrated representation of d_i. Here, we apply an RNN. Considering $\mathbf{I}_{d_i} = \{\mathbf{T}_{i,1}, \ldots, \mathbf{T}_{i,l_{d_i}}\}$, where $\mathbf{T}_{i,j}$ is the enhanced representation of the j-th term in d_i, the hidden state $\mathbf{h}_t$ of the RNN is described as:

$$\mathbf{h}_{i,t} = \tanh(\mathbf{W}_i[\mathbf{h}_{i,t-1}; \mathbf{T}_{i,t}] + \mathbf{b}_i), \quad t \in [1, l_{d_i}]. \tag{7}$$

The last hidden state $\mathbf{h}_{i,l_{d_i}}$ is used as the integrated representation of the document d_i. To simplify the notation, we use $\mathbf{v}_i = \mathbf{h}_{i,l_{d_i}}$, and it contains matching signals between the selected document d_i and the candidate document d.

Afterwards, when obtaining integrated representations of all selected documents, we employ another RNN to aggregate the information of the whole selected documents sequence as:

$$\mathbf{h}_{d,k} = \tanh(\mathbf{W}_d[\mathbf{h}_{d,k-1}; \mathbf{v}_{k-1}] + \mathbf{b}_d), \quad k \in [1, |\mathcal{C}|]. \tag{8}$$

We use the last hidden state $\mathbf{h}_{d,|\mathcal{C}|}$ to represent the selected document sequence. This vector contains the matching information between the candidate document and the selected document sequence, and it is denoted as $\mathbf{v}_d$ for simplification. Note that in practice, we use GRU cells for all RNNs.

Ranking Score. Inheriting the spirit of MMR [4], the final ranking score is calculated based on both the relevance and the novelty. For a candidate document d, its ranking score s is calculated as:

$$s = \text{MLP}(\text{ReLU}(\text{MLP}([\mathbf{x}_d; \mathbf{v}_d]))), \tag{9}$$

where $\text{MLP}(\cdot)$ is a multi-layer perceptron, ReLU is ReLU activation function, and ; is concatenation operation. $\mathbf{x}_d$ is a group of relevance features of d regarding the query q. Following previous studies [17,21,22], we use some traditional IR features, such as BM25 and TF-IDF, to measure the relevance. For each ranking position, our model greedily selects the best document with the highest score s. When a document is selected, it will be added to $\mathcal{C}$. This process will be repeated until all the documents are selected.

3.3   Model Training and Inference

Loss Function. In the process of training, we use the sum of all the documents' ranking score s_i as the score s_r of a given ranking sequence r. Following previous work [17,22], we apply a list-pairwise sampling approach to generate training samples in limited datasets. With the positive and negative ranking pair (r_1, r_2), the loss function for list-pairwise samples is defined as a binary classification log-loss formation:

$$\mathcal{L} = \sum_{q \in Q} \sum_{s \in S_q} |\Delta \mathcal{M}| [y_s \log(P(r_1, r_2)) + (1 - y_s) \log(1 - P(r_1, r_2))]. \qquad (10)$$

Here $|\Delta \mathcal{M}| = |\mathcal{M}(r_1) - \mathcal{M}(r_2)|$, and $P(r_1, r_2) = \sigma(s_{r_1} - s_{r_2})$, where $\sigma(\cdot)$ is the sigmoid function. Due to space limitation, we omit the detailed introduction of list-pairwise sampling method, and more details can be found in [17]. MatchingDIV is optimized in an end-to-end manner, where the BERT encoder is fine-tuned, and the other components are trained from scratch.

Reducing Online Inference Latency. As MatchingDIV leverages BERT, which is a large model, to encode the documents, for online ranking tasks, we propose two strategies to reduce the ranking latency.

(1) *Late-interaction Strategy.* The original design of BERT suggests to concatenate two documents as a long sequence and model their relationship through the first special token. However, in our task, it is impractical to concatenate each selected document with the candidate document in online scenario due to the high computation cost. Therefore, Following the previous work [18], we apply late-interaction strategy to decoupling the encoding and interaction of the documents so that the document representation can be pre-computed and stored offline. As a result, the online inference latency of computing the document representations can be omitted.

(2) *Ranking-top Strategy.* In MatchingDIV, the computational cost is increasing with the length growth of selected document sequence $\mathcal{C}$. Hence, we propose a ranking-top strategy to reduce the computational cost of online document interactions. For an initial ranking list $\mathcal{D}$ with m documents, MatchingDIV takes all the m candidate documents as input and return m ranking scores. Then, MatchingDIV greedily selects the best document and iteratively adds it to the ranking sequence $\mathcal{R}$. When $|\mathcal{R}|$ grows to the maximum number n ($n < m$), the ranking process will stop early, and all remained candidate documents in $\mathcal{D}$ will be directly appended into $|\mathcal{R}|$. In other words, with the ranking-top strategy, only the top N documents in $\mathcal{R}$ are re-ranked in diversity. The computational cost of document interactions between selected and candidate documents can thus being reduced. In practical application, search result diversification aims to satisfy user intents in former ranking positions, so it is unnecessary to spend much time for the latter positions.

4   Experiment Settings

4.1   Datasets and Metrics

We use the Web Track dataset from TREC 2009 to 2012 with 198 queries in total. The query #95 and #100 without diversity judgements are not used. Each query includes 3 to 8 user intent annotations, and the relevance rating is marked as relevant or irrelevant at intent level. We use the preprocessed relevance feature data provided by Jiang et al. [17] on GitHub[1]. The data include 18 relevance features for each query and subquery generated by traditional IR models. More details about those features can be found in [17]. The title (if available) and content are concatenated together for tokenization, and we only use the first $l_d = 80$ terms for the document since the documents are usually too long for document interaction.

The evaluation metrics in our experiments are the official Web Track diversity metrics, including α-nDCG [7], ERR-IA [5], and NRBP [2]. Similar to previous work [17,30,31,36], we also apply the metrics of Precision-IA [1] (denoted as Pre-IA) and Subtopic Recalls [34] (denoted as S-rec). We use the top-50 documents of Indri initial rankings as inputs, and all those metrics are computed on the top 20 results of the diversified ranking lists. Two-tailed paired t-test are used to conduct significance testing with p-value <0.05. In the significance testing, MatchingDIV is compared with PAMM-NTN as the state-of-the-art supervised implicit model and DSSA as the best explicit model.

4.2   Model Settings

In the training phase, we use 5-fold cross validation to tune the parameters in all experiments with the widely used α-nDCG@20. In each fold, there are 160 queries for training and 40 queries for testing. In our experiment, the hidden size of GRU is 128, and the BERT-based embeddings are compressed into 128 dimension. The batch size is 32. We use Adam [19] optimizer. The learning rate of the BERT encoder is $3e-5$, while that of other network components is $1e-3$.

We compare MatchingDIV with baselines including:

(1) Non-diversified approaches: **Lemur**, **ListMLE**. These two ad-hoc ranking methods doe not consider diversity.
(2) Explicit diversification methods: **xQuAD** [24], **PM2** [9], **TxQuAD**, **TPM2** [10], **HxQuAD**, **HPM2** [14]. These are representative unsupervised explicit methods. **DSSA** [17] is a supervised method, which models the diversity of the documents with subtopic attention using RNNs. This is the state-of-the-art explicit diversification methods. Note that our method uses BERT as the document encoder, and we also equip DSSA with BERT and denote this variant as **DSSA (BERT)** for a fair comparison.
(3) Implicit diversification methods: **R-LTR** [36], **PAMM** [30], **NTN** [31]. They are representative supervised implicit methods. The neural tensor network (NTN) is used on both R-LTR and PAMM, denoted as **R-LTR-NTN** and **PAMM-NTN**, respectively.
(4) Ensemble methods: **DESA** [22] and **DVGAN** [21]. They are two ensemble methods that use both explicit (subtopic) features and implicit (document similarity) features.

---

[1] https://github.com/jzbjyb/DSSA.

5   Experimental Results

5.1   Overall Results and Discussion

Table 1 shows the results of all models. We can observe: (1) MatchingDIV out-performs all the implicit and explicit baseline models, and the improvement is statistically significant (with p-value <0.05) on all the metrics except for Pre-IA. These results clearly demonstrate the effectiveness of our proposed RMEDiv. (2) Compared with those approaches based on pre-trained document embeddings, our framework can capture and aggregate the fine-grained matching signals between selected and candidate documents, thus improving the performance of search result diversification. (3) Intriguingly, MatchingDIV, as an implicit method without using subtopic coverage, can perform slightly better than the ensemble approaches (DVGAN and DESA). This reflects that our proposed interaction-based document matching is very effective. Besides, this result also implies the advantage of enhancing the relevance matching component for diversification. (4) Pre-trained language models (such as BERT) are reported to have great capability of representation. By integrating it into the baseline DSSA, we see a slight performance improvement. However, there is still a large gap between

Table 1. Performance of all approaches. The baselines include: (1) non-diversed methods; (2) explicit methods; (3) implicit methods; and (4) ensemble methods. The best results are in bold. † indicates that our model significantly outperforms all implicit and explicit approaches (p-value <0.05 in two-tailed paired t-test).

Methods	ERR-IA	α-nDCG	NRBP	Pre-IA	S-rec
(1) Lemur	.271	.369	.232	.153	.621
(1) ListMLE	.287	.387	.249	.157	.619
(2) xQuAD	.317	.413	.284	.161	.622
(2) TxQuAD	.308	.410	.272	.155	.634
(2) HxQuAD	.326	.421	.294	.158	.629
(2) PM2	.306	.411	.267	.169	.643
(2) TPM2	.291	.399	.250	.161	.639
(2) HPM2	.317	.420	.279	.172	.645
(2) DSSA (doc2vec)	.350	.452	.318	.184	.645
(2) DSSA (BERT)	.352	.457	.319	.181	.656
(3) R-LTR	.303	.403	.267	.164	.631
(3) PAMM	.309	.411	.271	.168	.643
(3) R-LTR-NTN	.312	.415	.272	.166	.644
(3) PAMM-NTN	.311	.417	.272	.170	.648
(3) **MatchingDIV (Ours)**	**.366**†	**.467**†	**.334**†	**.185**	**.659**†
(4) DVGAN	.367	.465	.334	.175	.660
(4) DESA	.363	.464	.332	.184	.653

the performance of DSSA (BERT) and that of our proposed MatchingDIV. This indicates that the better performance we obtained is not merely due to BERT embeddings but also to the proposed interaction-based document matching.

Table 2. Effects of different pretrained models

Settings	ERR-IA	α-nDCG	NRBP	Pre-IA	S-rec
Electra-base-dicsriminator	**.366**	**.467**	**.334**	**.185**	**.659**
Distilbert-base-uncased	.360	.463	.329	.182	.655
Bert-base-uncased	.363	.465	.332	.184	.659

5.2 Effect of Different Encoder

We further investigate the effect of different model settings in MatchingDIV. Specifically, we try other pretrained models provided by Huggingface[2] as document encoder. The following models are tested: the basic BERT model "bert-base-uncased", the DistilBERT [23] model "distilbert-base-uncased", and the ELECTRA [6] model "electra-base-discriminator". Results are shown in Table 2. The results show that the BERT-base model performs slightly better than DistilBERT, while the ELECTRA has achieved the best performance. It is worth noting that all these variants achieve better performance than existing baseline methods. This further validates the effectiveness of our proposed interaction-based document matching method.

Table 3. Results of average online ranking time per query.

Setting	Average time online (ms)	α-nDCG@20
$n = 5$	19	.458
$n = 10$	45	.464
$n = 20$	113	.466
$n = 50$	295	.466

5.3 Inference Latency for Online Ranking

As we introduced in Sect. 3.3, the inference latency is very important when applying a diversification model in practice. On the one hand, MatchingDIV employs

---

[2] https://github.com/huggingface/transformers.

the late-interaction mechanism, allowing to encode the documents offline. Therefore, the computational time of encoding the documents into term-level embeddings can be omitted. On the other hand, with the ranking-top strategy, MatchingDIV only generates the top n documents of the diversified ranking list $\mathcal{R}$. To investigate the effect of this strategy, we set $n = \{5, 10, 20, 50\}$ and test the model's inference time and corresponding performance in terms of α-nDCG@20. Experimental results are shown in Table 3.

From these results, we can find that the average online inference time growth approximate linearly regarding the increasing size of n. The performance also improves from $n = 5$ to $n = 20$. After that, when $n = 50$, the average inference time is increasing, but the performance is no longer improved. This is because the evaluation metric α-nDCG@20 only considers the diversification in top 20 documents. Indeed, this result is consistent with the goal of search result diversification. The diverse ranking model aims at satisfying user intents at former ranking positions rather than spending lots of time on the latter ranking positions. When $n = 20$, the performance is good, and the online ranking latency is 113 ms, demonstrating that our framework is effective and efficient.

6   Conclusion and Future Work

In this work, we proposed an supervised framework MatchingDIV for integrating interaction-based document matching in implicit search result diversification. Based on BERT-based term embeddings of each document, MatchingDIV used cross-attention and GRUs to capture and aggregate low-level matching signals between selected documents and candidate documents. Compared with previous work, we are among the first to capture the fine-grained term-level matching signals for document selection in search result diversification. Experiment results showed that our framework can significantly outperform the previous SOTA explicit and implicit diversification method, and even outperform the ensemble diversification framework. Furthermore, with late-interaction and our proposed ranking-top strategy, the online ranking latency is affordable for actual search engines. These results demonstrate the advantage of employing interaction-based document matching for diversification tasks. As future work, we plan to also integrate query-document interactions, which may bring further improvement for diverse ranking tasks.

Acknowledgments. This work was supported by National Natural Science Foundation of China No. 61872370 and No. 61832017, and Beijing Outstanding Young Scientist Program No. BJJWZYJH012019100020098. We thank all the anonymous reviewers for their insightful comments.

References

1. Agrawal, R., Gollapudi, S., Halverson, A., Ieong, S.: Diversifying search results. In: WSDM (2009)

2. Baeza-Yates, R., Hurtado, C., Mendoza, M.: Query recommendation using query logs in search engines. In: Lindner, W., Mesiti, M., Türker, C., Tzitzikas, Y., Vakali, A.I. (eds.) EDBT 2004. LNCS, vol. 3268, pp. 588–596. Springer, Heidelberg (2004). https://doi.org/10.1007/978-3-540-30192-9_58
3. Blei, D.M., Ng, A.Y., Jordan, M.I.: Latent Dirichlet allocation. J. Mach. Learn. Res. **3**, 993–1022 (2003)
4. Carbonell, J.G., Goldstein, J.: The use of MMR, diversity-based reranking for reordering documents and producing summaries. In: SIGIR (1998)
5. Chapelle, O., Metlzer, D., Zhang, Y., Grinspan, P.: Expected reciprocal rank for graded relevance. In: CIKM. ACM (2009)
6. Clark, K., Luong, M., Le, Q.V., Manning, C.D.: ELECTRA: pre-training text encoders as discriminators rather than generators. In: ICLR. OpenReview.net (2020)
7. Clarke, C.L.A., et al.: Novelty and diversity in information retrieval evaluation. In: SIGIR. ACM (2008)
8. Dai, Z., Xiong, C., Callan, J., Liu, Z.: Convolutional neural networks for soft-matching n-grams in ad-hoc search. In: WSDM. ACM (2018)
9. Dang, V., Croft, W.B.: Diversity by proportionality: an election-based approach to search result diversification. In: SIGIR (2012)
10. Dang, V., Croft, W.B.: Term level search result diversification. In: SIGIR (2013)
11. Devlin, J., Chang, M., Lee, K., Toutanova, K.: BERT: pre-training of deep bidirectional transformers for language understanding. In: NAACL-HLT. Association for Computational Linguistics (2019)
12. Dou, Z., Song, R., Wen, J.: A large-scale evaluation and analysis of personalized search strategies. In: WWW. ACM (2007)
13. Guo, J., et al.: A deep look into neural ranking models for information retrieval. Inf. Process. Manag. **57**(6), 102067 (2020)
14. Hu, S., Dou, Z., Wang, X., Sakai, T., Wen, J.: Search result diversification based on hierarchical intents. In: CIKM (2015)
15. Huang, P., He, X., Gao, J., Deng, L., Acero, A., Heck, L.P.: Learning deep structured semantic models for web search using clickthrough data. In: CIKM. ACM (2013)
16. Jansen, B.J., Spink, A., Saracevic, T.: Real life, real users, and real needs: a study and analysis of user queries on the web. Inf. Process. Manag. **36**(2), 207–227 (2000)
17. Jiang, Z., Wen, J., Dou, Z., Zhao, W.X., Nie, J., Yue, M.: Learning to diversify search results via subtopic attention. In: SIGIR (2017)
18. Khattab, O., Zaharia, M.: ColBERT: efficient and effective passage search via contextualized late interaction over BERT. In: SIGIR. ACM (2020)
19. Kingma, D.P., Ba, J.: Adam: a method for stochastic optimization. In: Bengio, Y., LeCun, Y. (eds.) ICLR (2015)
20. Le, Q., Mikolov, T.: Distributed representations of sentences and documents. In: ICML 2014 (2014)
21. Liu, J., Dou, Z., Wang, X., Lu, S., Wen, J.: DVGAN: a minimax game for search result diversification combining explicit and implicit features. In: SIGIR (2020)
22. Qin, X., Dou, Z., Wen, J.: Diversifying search results using self-attention network. In: CIKM. ACM (2020)
23. Sanh, V., Debut, L., Chaumond, J., Wolf, T.: Distilbert, a distilled version of BERT: smaller, faster, cheaper and lighter. CoRR abs/1910.01108 (2019)
24. Santos, R.L.T., Macdonald, C., Ounis, I.: Exploiting query reformulations for web search result diversification. In: WWW (2010)

25. Shen, Y., He, X., Gao, J., Deng, L., Mesnil, G.: Learning semantic representations using convolutional neural networks for web search. In: WWW. ACM (2014)
26. Silverstein, C., Henzinger, M.R., Marais, H., Moricz, M.: Analysis of a very large web search engine query log. In: SIGIR Forum, vol. 33, no. 1 (1999)
27. Song, R., Luo, Z., Wen, J., Yu, Y., Hon, H.: Identifying ambiguous queries in web search. In: WWW. ACM (2007)
28. Tao, C., Wu, W., Xu, C., Hu, W., Zhao, D., Yan, R.: Multi-representation fusion network for multi-turn response selection in retrieval-based chatbots. In: WSDM (2019)
29. Vaswani, A., et al.: Attention is all you need. In: NeurIPS (2017)
30. Xia, L., Xu, J., Lan, Y., Guo, J., Cheng, X.: Learning maximal marginal relevance model via directly optimizing diversity evaluation measures. In: SIGIR (2015)
31. Xia, L., Xu, J., Lan, Y., Guo, J., Cheng, X.: Modeling document novelty with neural tensor network for search result diversification. In: SIGIR (2016)
32. Xiong, C., Dai, Z., Callan, J., Liu, Z., Power, R.: End-to-end neural ad-hoc ranking with kernel pooling. In: SIGIR. ACM (2017)
33. Yue, Y., Joachims, T.: Predicting diverse subsets using structural SVMs. In: ICML. ACM International Conference Proceeding Series, vol. 307 (2008)
34. Zhai, C., Cohen, W.W., Lafferty, J.D.: Beyond independent relevance: methods and evaluation metrics for subtopic retrieval. In: SIGIR (2003)
35. Zhou, X., et al.: Multi-turn response selection for chatbots with deep attention matching network. In: ACL (2018)
36. Zhu, Y., Lan, Y., Guo, J., Cheng, X., Niu, S.: Learning for search result diversification. In: SIGIR (2014)

Various Legal Factors Extraction Based on Machine Reading Comprehension

Beichen Wang[1], Ziyue Wang[1(✉)], Baoxin Wang[1], Dayong Wu[1],
Zhigang Chen[1], Shijin Wang[1,2], and Guoping Hu[1]

[1] State Key Laboratory of Cognitive Intelligence, iFLYTEK Research, Beijing, China
{zywang27,bxwang2,dywu2,zgchen,sjwang3,gphu}@iflytek.com
[2] iFLYTEK AI Research (Hebei), Langfang, China

Abstract. With the rapid growth of legal cases, professionals are under pressure to go through lengthy documents and grasp informative pieces of text in the limited time. Most of the existing techniques focus on simple legal information retrieval task, such as name or address of the prosecutor or the defendant, which can be easily accomplished with the help of handcrafted patterns or sequence labeling methods. Yet complicated texts always challenge such pattern-based methods and sequence labeling approaches. These texts state the same facts or describe the same events, but they do not share common or similar patterns. In this paper, we design a unified framework to extract legal information in various formats, including directly extracted information (a piece of span) and information that needs to be deduced. The framework follows the methodology to answer questions in machine reading comprehension (MRC) tasks. We treat the extraction fact labels as the counterpart of questions in MRC task and propose several strategies to represent them. We construct several datasets regarding different cases for training and testing. Our best strategy achieves up to 4% enhancement in F1 score on each dataset compared to the MRC baseline.

Keywords: Information extraction · Machine reading comprehension · Legal information

1 Introduction

Nowadays, driven by increasingly complicated legal provisions and cases, both ordinary parties and legal workers are eager to use technical means to assist in analysis. In the process of assisting judicial work, it is an indispensable ability to extract various forms of required information efficiently and correctly. The information are sometimes pieces of text in the document, such as event descriptions, actions and entity names, or conclusions that are not directly stated and need deducing from the original text. Such information are usually crucial to the final sentence, which are called *legal factors* in the legal industry. Legal factors are closely cohered with legal case types and each type has a fixed factor list. Given the case type, judges or other legal workers are clear of what information to seek

© Springer Nature Switzerland AG 2021
H. Lin et al. (Eds.): CCIR 2021, LNCS 13026, pp. 16–31, 2021.
https://doi.org/10.1007/978-3-030-88189-4_2

from the document. However, different forms of legal factors cannot be easily extracted by a single model simultaneously using existing techniques. Thus, we hope to build a unified framework to extract all factors regarding the case type.

Generally, information extraction task [6] extracts entities (person, organization, etc.) and facts (relations, events, etc.) from given texts [16], helping to acquire the desired information and reconstruct massive contents. This is usually done by sequence labeling models, such as Lattice LSTM [18] and transformer-based models [9,17]. They conduct experiments on Chinese named entity recognition (NER) task, and achieve high performance in benchmarks, such as Chinese NER MSRA [8], OntoNotes 4.0, and Resume NER [18]. However, they fail to maintain competitive results in our task. Our extraction targets are of multiple forms, including spans and deductions. Spans are often entity names and event descriptions, while deductions include answers to predefined legal conditions, for example, whether the document agrees (answers "Yes") or opposes (answers "No") to a given condition. These challenge the sequence labeling models since the answers are not spans from the context. Inspired by MRC tasks [7,12,13], whose question types are similar to ours, for example, asking for a named entity (span) and deducing a piece of opinion ("Yes" or "No"), we can employ such ideas to solve the legal factor extraction task. The difference lies in that the questions of MRC tasks are arbitrary, but the extraction fact labels (the counterpart to questions in MRC) are fixed along with the case type. We pre-define these fixed legal factors to a given case as a set of *extraction fact labels*.

In this paper, we design a unified framework, particularly for multi-formed legal factor extraction. It follows the MRC methodology to encode and interact with the document and the extraction fact labels. Some researchers proved that more specified and informative queries could improve the extraction precision [10]. As our extraction fact labels are fixed, to achieve better performance, we expand the labels to obtain more specified expressions with query expansion components before the extraction. Our framework solves the problem discussed above and our best strategy achieves 4% improvement compared to the baseline. Our contributions are as follows: i. We achieve automatic extraction of various forms of legal factor with a single model. ii. We proposed a strategy to represent legal factors that outperform the others.

2   Related Works

Some existing works investigates the feasibility of using MRC approaches to solving information extraction tasks. Li et al. [10] use the MRC model to solve the named entity recognition task and achieved good results in nested entity recognition. For each more nested entity, one more question requires to be answered. And the reading comprehension model is designed to handle the question-answering task. They suppose that the reading comprehension model is a natural solution to the nested entity problem. Similarly, in our task, information of factors is often nested or overlapped, we could exploit the MRC framework for legal factor extraction.

Query expansion is comprehensively used because of its simplicity and practicality. There are several ways to reach query expansion. The IBM algorithm in the machine translation model is migrated to directly rewrite queries [5]. People find that it is too rough to directly rewrite queries, and instead use the Seq2Seq model to incorporate richer semantic information, and reinforcement learning is used to fine-tune the rewrite model [1]. Of course, there are other ideas, such as using a large amount of query click data, various concept words are mined out. Further, associating the concept words with the knowledge graph in order to replace the simple query clustering scheme [11].

3   The Legal Factor Extraction Framework

3.1   Architecture of the Framework

Primarily, we introduce two important concepts, *legal factors* and the *legal factor extraction* task. As mentioned in Sect. 1, legal factor is a jurisprudence concept, referring to the key information that affects the final sentencing, which includes the descriptions of a judicial event, such as persons, actions, causes, consequences and etc., and the conclusions to predefined judicial conditions. We define the legal factor extraction task as the procedure to find out the text stating such information from original paragraphs. Our framework consists of three parts: input documents and extraction fact labels embeddings, documents and labels interactive encoding, and result answers prediction. The overall structure is shown in Fig. 1 and Fig. 2 is an example shows how our framework actually works.

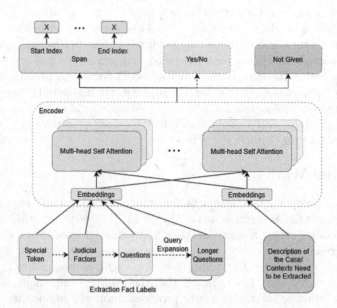

Fig. 1. Architecture of the framework. All the four forms of extraction fact labels can be fed into the encoder independently for extraction. The dotted arrows refer to the semantic enhancement.

In the embedding part, the inputs are the case descriptions and extraction targets. Note that the case type is an inherent attribute of a case description and the targets of extraction are determined along with the case type. We design different representations of the extraction fact labels and gradually enriched their semantic. First, we employ special tokens ([Token1], [Token2], etc.) as the targets. These tokens function as signals prompting the model to extracting different legal factors. Then, we represent the labels by text of legal factors, such as "伤情等级" (the injury level) and "经过/手段" (course/means). Further, the legal factors are expanded into more specific questions, for instance, "被害人伤情等级?" (What is the victim's injury level?) and "被告人是否使用工具?" (If the defendant use tools?). Finally, the questions are reinforced to form longer questions via query expansion methods. We will discuss different extraction targets in the following subsection. The rest parts interactively encode documents and extraction fact labels with multi-head self attention [15], and predict the results according to the types of extraction fact labels. If it is unanswerable, an empty string returned. Otherwise, it answers "Yes" or "No" to a yes/no question, or returns the result of the extraction through the start and end indexes in the input passage.

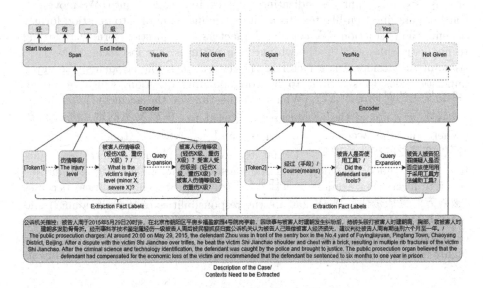

Fig. 2. Examples of the workflow (Left: Answer as span. Right: Answer as "YES"). We do not translate sentences after data augmentation (DA) because DA is usually a paraphrase of the original question, and the English version could remain unchanged.

3.2 Strategies for Representing Extraction Fact Labels

The Special Tokens. To verify our hypothesis that the different forms of legal factor can be extracted by a unified framework, we represent each extraction fact label as different [Token] rather than using the actual text of factor. They act as signals guiding the model to extract designated content. Specific, we set a series of special tokens according to the legal factors that need to be extracted, where the factors are one-to-one mapped to the tokens. It is equivalent to cipher or semaphore. When the model encounters a special token, it learns to extract a specific kind of information. Commencing from the special tokens, we enrich the semantic gradually in the rest extraction fact labels representations.

From Factor to Question. Our produced datasets are dedicated to the type of cases, and the information that needs to be extracted from each dataset is summarized according to the experiences of the judiciary (legal factors). Taking the dataset of traffic accidents as an example, the factors are "事故时间" (time of the accident), "事故地点" (site of the accident), "责任划分" (division of responsibilities), "造成损失" (the loss), etc. These factors are the counterpart of questions in typical question-answering datasets such as SQuAD [13], SQuAD2.0 [12], and CJRC [4]. Although the factors are not in the form of interrogative sentences, they contain all the core information needed for extraction. Hence, we use it the factors as another form of extraction fact labels.

Nevertheless, the factors are not written in general question format, they are just short key phrases indicating what content to look into. We move forward to gain question-like text to see if expanding a factor into question format benefits the extraction. We convert the factors into question format according to linguistic rules such as the wh-question patterns. For instance, "事故时间" (time of the accident) is a factor in the dataset of traffic accidents. We apply linguistic rules to switch it to its corresponding question, which results in "事故发生的时间是什么?" (What was the time of the accident?). Similarly, "事故地点" (site of the accident) is switched to "事故发生的地点是什么?" (Where did the accident take place?), and "造成损失" (the loss) turns to "原告方遭受的损失有哪些?" (What losses does the plaintiff suffer?). Some factors are expanded into more than one question because they refer to multiple concrete facets. For example, "责任划分" (division of responsibilities) becomes "被告方的责任是什么?" (What are the responsibilities of the defendant?) and "原告方的责任是什么?" (What are the responsibilities of the plaintiff?). Such expansion literally enriches the semantic compared to the vanilla factors, and is highly likely to enhance the performance in the question-answering experiment.

From Question to Questions. Following the outcome (converted question) of last section, we use query expansion to further enrich the semantic information of the extraction factor labels. In this section, each extraction factor label is represented by longer or multiple questions. The expansion is quite simple and straightforward, namely synonym expansion. Specifically, there are two ways to expand a sentence with its synonym:

- The first is word-wise synonym expansion. A sentence is first segmented into words using LTP [2] as the word segmentation module. Then the corresponding synonyms of each word are directly concatenated following the original word order. We design two strategies to find the synonyms of each word, "M_1" and "M_2". "M_1" looks up the existing dictionary for synonyms, such as BigCilin [19]. While "M_2" builds a special dictionary concerning both contexts and questions in each dataset. The embeddings of words in the whole created dictionary are generated by BERT [3]. We use the embedding vectors for calculating cosine similarities of the input word and the rest words from the dictionary. We rank the candidates and resume top K words as retrieved synonyms. Then, the synonyms are directly piled up to form a new sentence. For example, in the dataset of recourse for labor remuneration when supposing $K = 3$, the question "当事人 是否 签订 劳动 合同?" (Whether the party signs the labor contract?) will be expanded as "当事人原告被告 是否否应该 签订签定签署 劳动劳动者劳动力 合同合约合同规定?".

- The second is sentence-wise synonym expansion and we name it data augmentation (DA). The sentence is treated as a whole when searching for synonyms. To augmentate a whole sentence, we need to gain its the general semantic representation. Hence, we rely on the corpus of Baidu zhidao, which is a knowledge encyclopedia written in Chinese. This corpus is used to pretrain language model that better understand the correlation between Chinese words and phrases. We refer readers to SimBERT [14] for training details. We use this model to gain a ranked list of K similar sentences. After acquiring the sentence-wise synonyms, we propose several strategies to join them. In addition to direct concatenation, we utilize bi-LSTM and bi-GRU to encode semantic of these synonyms. Compared to the direct concatenation, this strategy extract hidden features and control dimension of the output in order to satisfy BERT's input length limit. The direct concatenation easily exceeds this limitation and the rest part of text is ignored. We report selected augmentation results of $K = 5$ in Table 1 and Table 5 (in the appendix).

Table 1. Examples of data augmentation of recourse for labor remuneration (RLR) (See Appendix for other datasets). We report two typical questions for each dataset. Those without translations share the same English version with the original question's.

Cases	Original Question	Data Augmentation (DA) (K=5)
RLR	劳动者的基本工资数额是多少? (What is the basic salary of workers?)	劳动者工资的基本数额是多少?
		劳动者的基本工资是多少?
		劳动者每月工资基数是多少?
		劳动者每月基本工资是多少?
		劳动者工资的基本数额和年终奖是多少? (How much are the basic salary and bonus?)
	离职的原因是什么? (What's the reason for quitting?)	离职的主要原因是什么?
		离职是什么原因造成的? (What causes the quitting?)
		辞职的原因是什么?
		员工离职的原因是什么?
		辞职的原因是什么?

4 Experiments

4.1 Datasets and Experimental Settings

We composed three datasets regarding the following case types, intentional injury (II), recourse for labor remuneration (RLR), and refusal to execute judgments or rulings (REJR). The statistics are shown in Table 2. We select 200 question-answer pairs from each dataset as the testset. Note that the number of questions answered by "No" is markedly low. This is a common feature of judicial data, that officers tend to ask what they've already had evidence by hands, leading to few negative answers. We do not manually disturb this, because we believe that it is better to restore the actual dataset distribution.

As for experimental settings, we employ BERT-base encode the documents and extraction fact labels and refer readers to BERT [3] for detailed model descriptions and hyper-parameter settings. Since the input length of Bert is limited by 512, the length (L_F) allocated to the extraction fact label should also be constrained and well-designed. Overall, the input extraction fact label tokens of our model is no longer than 60, and the rest are preserved for document input. Specific, $L_F = 1$ if using [Token] as the representation; $L_F \leq 10$ for factors and questions; while $L_F \leq 60$ for expansions. In practice, most of the expanded questions are longer than 10, so any $K > 5$ results in input length suppressing. We will seek the best effect by varying K.

Table 2. Statistics of datasets. # refers to the number. NG refers to "Not Given".

Cases	#Factors	#Document	#Questions	Answer types			
				Yes	No	Span	NG
RLR	17	992	5628	1233	189	8422	946
II	13	974	9840	695	246	7840	1059
REJR	13	960	8320	2017	25	4142	2158

4.2 Experiments of Different Methods for Legal Factor Extraction

Before carrying out experiments with MRC models, we try the sequence labeling methods to address this task. We convert our legal factor extraction datasets into the sequence labeling format, where the case descriptions in our datasets are labeled with BIOES (B-begin, I-inside, O-outside, E-end, S-single) tags. Further, we reproduce two entity recognition algorithms, Lattice LSTM [18] and Chinese BERT-base [3]. Because these models require to find span text from the given passage, only questions answered by span are preserved and the corresponding questions turn out to be the tags. The number of such tags are limited, so that we could number as Q1, Q2, etc. Specifically, for the question Q1, we find the corresponding answer span in the context and label the tokens within the span as "B-Q1 I-Q1 I-Q1 ... E-Q1". The sequence labeling experiments are compared to our proposed question-answering approaches in Table 3. Unfortunately, the results do not live up to expectations, which we will discuss in Sect. 5.

4.3   The Structure of Questions and the Design of Query Expansion

We investigate different methods to figure out the query expansion module that works the best and design a serious of ablation studies. In this section, we introduce these methods in details:

- "Q_{ori}": To Directly generate questions from the factors. The result of the question-answering experiment in "Q_{ori}" will be the baseline of the whole experiments.
- "$modify$": We manually modify some questions, mainly to fix some mistakes caused by the generation process and to make them closer to the contexts. The modification includes correction of typos and examination of punctuation characters. In the II dataset, we modify the question "受害人的受害经过是怎样的?" (What happened to the victim?) to "被害人的被害经过是怎样的?". Although the English translation for both "受害人" and "被害人" is victim, only "被害人" appears in the contexts of the dataset rather than "受害人". Therefore, we modify this word to be consistent with the context.
- "$+F$": To directly concatenate the factors with original questions as a whole.
- "QE_{syn}-$M1$": Word-wise expansion using method $M1$ described in Sect. 3.2.
- "QE_{syn}-M_2": Word-wise expansion using method $M2$ described in Sect. 3.2.
- "QE_{da}-$random$": To replace the original question with a randomly question obtained by data augmentation.
- "QE_{da}-$lstm$": To add a bi-LSTM layer after data augmentation to extract semantic information from augmented questions.
- "QE_{da}-gru": To add a bi-GRU layer after data augmentation to extract semantic information from augmented questions.
- "QE_{da}-$topK$": To directly concatenate the first K augmented sentences with the original question. We experiment with $K = 1, 3, 5$.
- "QE_{da}-mix": To only expand questions whose answers are spans, but use the original question for those answered by "Yes" or "No", and those with no answers (NG).
- "QE_{da}-yn": Do not expand question those answered by "Yes" or "No", but perform expansion on other kinds of questions.
- "QE_{da}-$last5$": To concatenate the last five out of ten augmented sentences with the original question. We hypothesize that taking the last five sentences increases the diversity of semantic because they are less similar to the original question than the first five sentences.

Among, the above strategies, QE_{da}-$topK$, QE_{da}-mix, QE_{da}-yn and QE_{da}-$last5$ are four ablation studies that reveal what indeed leads to the enhancement of extraction results.

4.4   Results

The results of different methods to extract legal information and the performances of different forms of extraction fact labels are reported in Table 3. Not as expected, the sequence labeling methods much underperform the rest methods. The F1 score of Lattice only reaches 11.17%. Similarly, $Seq2Seq_{BERT}$ does not perform much better, with F1 score 22.8%. We can conclude that the sequence labeling methods are incapable of handling our task. Among the forms of extraction fact labels, the one with query expansion module achieves best F1 results for RLR (88.51), II (86.11) and REJE (81.47) datasets. This method also contributes to the highest EM scores on RLR (77.07) and REJR (67.34) datasets. While in the II dataset, representing fact labels as questions peaks by the EM score of 74.88. Table 3 shows that along with the increasing of semantic information from tokens to expanded questions, the results are broadly improved, reaching up to 20% in F1 and 15% in EM. These prove the idea that richer semantic features boosts the model's performance.

Table 4 and Table 6 (in the appendix) report the results of query expansion strategies and the ablation experiments. Taking RLR dataset as an example, we apply all the methods and strategies described in Sect. 4.3 to test the effectiveness. Among these strategies, the data augmentation-based strategies outperform the others. The synonym-based experiments show similar pattern as the RLR for the other two datasets. So we omit the synonym-based results due to the page limitation and focus on the data augmentation-based experiments. In RLR dataset, "QE_{da}-top5" has a positive effect and its EM and F1 enhance about 1.5% and 2%, respectively compared to "QE_{da}-lstm" and "QE_{da}-gru". After these experiments, we realize that query expansion methods work unequally for all kinds of questions. Some work well on questions whose answers are spans, but poorly on "Yes", "No", or "NG" questions. Hence, apart from expanding question of all answer types (QE_{da}-topK), we design another experiments, QE_{da}-mix, QE_{da}-yn, and QE_{da}-last5, as ablation experiments to test the influence on different types of extractions and to find out the best strategy. The QE_{da}-mix achieves an overall good result and is the recorded strategy in Table 3, but it affects datasets differently.

Table 3. Exact Match(%) and **F1**(%) of different forms of extraction fact labels (token, factor, question and expanded question) and sequence labeling method. Specific, F1$_{Seq1}$: the F1 of Lattice [18]; F1$_{Seq2}$: the F1 of $Seq2Seq_{BERT}$; Expansion: QE_{da}-mix.

Cases	Seq. labeling		Token		Factor		Question		Expansion	
	F1$_{Seq1}$	F1$_{Seq2}$	EM	F1	EM	F1	EM	F1	EM	F1
RLR	11.17	22.80	69.39	82.04	72.96	85.49	73.98	85.13	**77.04**	**88.51**
II	–	–	69.35	81.30	71.36	82.92	**74.88**	86.10	72.86	**86.21**
REJR	–	–	52.26	60.66	58.29	66.24	65.83	80.76	**67.34**	**81.47**

5　Analyses

5.1　Extract Legal Factor as a Human Judge

Table 3 examines the feasibility of sequence labeling approaches in solving the proposed legal factor extraction task and confirms that they are unsuitable to our task. The "entities" in our datasets are nearly ten times as long as the entities

Table 4. Results of special questions, reported in **Exact Match**(%), **F1** score(%), **Precision**(%) and **Recall**(%). Details of these methods are shown in Sect. 4.4

Cases	Methods	Total		Span		Not given		
		EM	F1	EM	F1	R	P	F1
Recourse for Labor Remuneration (RLR)	Q_{ori}	72.96	84.40	63.21	84.36	80.00	82.35	81.16
	$Q_{ori}+F$	70.92	83.98	57.55	81.70	82.86	80.56	81.69
	$Q_{ori}\text{-}modify$	73.98	85.38	61.32	82.40	85.71	83.33	84.50
	$QE_{syn}\text{-}M_1$	70.92	82.88	62.26	84.38	68.57	85.71	76.19
	$QE_{syn}\text{-}M_2$	71.94	84.36	59.43	82.39	80.00	80.00	80.00
	$QE_{syn}\text{-}M_2+F$	72.96	84.01	63.21	83.64	77.14	79.41	78.26
	$QE_{da}\text{-}random$	70.41	82.72	59.43	82.21	74.29	81.25	77.61
	$QE_{da}\text{-}top5$	75.51	86.25	**66.98**	**86.84**	80.00	87.50	83.58
	$QE_{da}\text{-}lstm$	71.43	84.39	58.49	82.46	82.86	78.38	80.56
	$QE_{da}\text{-}gru$	71.43	83.59	58.49	80.97	82.86	82.86	82.86
	$QE_{da}\text{-}mix$	**77.04**	**88.51**	63.21	84.41	100.00	92.11	**95.89**
	$QE_{da}\text{-}yn$	71.43	83.38	58.49	80.59	82.86	76.32	79.46
	$QE_{da}\text{-}top3$	71.43	82.91	59.43	80.66	82.86	74.36	78.38
	$QE_{da}\text{-}top1$	73.47	84.70	63.21	83.97	80.00	80.00	80.00
	$QE_{da}\text{-}last5$	72.96	84.73	62.26	84.04	80.00	82.35	81.16
Intentional Injury (II)	Q_{ori}	72.36	84.78	68.26	83.06	90.00	85.71	**87.80**
	$Q_{ori}\text{-}modify$	72.36	85.08	68.26	83.42	90.00	85.71	**87.80**
	$QE_{da}\text{-}top5$	71.86	83.56	68.26	82.21	85.00	89.47	87.18
	$QE_{da}\text{-}lstm$	70.85	83.70	67.66	82.98	80.00	84.21	82.05
	$QE_{da}\text{-}gru$	70.35	82.68	65.87	80.56	90.00	85.71	**87.80**
	$QE_{da}\text{-}mix$	**72.86**	**86.21**	**69.46**	**85.36**	85.00	89.47	87.18
Refusal to Execute Judgments or Rulings (REJR)	Q_{ori}	66.33	80.23	50.00	77.11	75.00	82.98	78.79
	$Q_{ori}\text{-}modify$	**67.34**	81.22	50.00	77.09	78.85	87.23	82.83
	$QE_{da}\text{-}top5$	**67.34**	81.19	50.98	**78.00**	76.92	86.96	81.63
	$QE_{da}\text{-}lstm$	65.83	80.01	49.02	76.69	78.85	82.00	80.39
	$QE_{da}\text{-}gru$	**67.34**	80.67	**51.96**	77.96	75.00	88.64	81.25
	$QE_{da}\text{-}mix$	66.83	**81.47**	49.02	77.57	78.85	91.11	**84.54**

in the standard NER datasets, such as Resume NER [18]. Our "entities" are essentially the answer spans to pre-defined questions, and are uneven in length. Whilst in the other datasets, most of the entities are of just a few tokens, and the number of entity tags are far less than ours. Therefore, the sequence labeling method is unqualified for information extraction task of this magnitude. These urges us to introduce the MRC-based means as better solutions.

In general MRC task, the model knows what answers to searching for because the questions are written in plain text. But in the "Token" experiment, we only feed the token symbols to the model without the specific target of the extraction descriptions. By doing so, we imitate the procedure of a judge reading through the documents. As we discussed in Sect. 1, these legal factors are fixed regarding the case type, and a judge can fast decide the what to looking for when given a case description. This experiment proves that the model can learn the semantic meanings through this special token without been told the exact target. To a certain extent, it completes the work of expert knowledge and artificial settings by itself. This differs our methods from general MRC approaches, and is instructive for follow-up experiments.

5.2   The Richer Semantic Information, the Higher Score

We can conclude from Table 3 that along with the increasing of semantic information from tokens to expanded questions, the performances are improved remarkably. Starting from the Token representation to the Factor and Question, and eventually the Expansion, the semantic meaning of the extraction fact labels get more specific and enhanced. Their corresponding results also get boosted gradually. In the most extreme example, the REJR dataset, the F1 score increases from about 60 to over 81, and the EM score grows about 15 from Token to Expansion. As shown in Table 3, there are surely information growth from Token to Factor and from Question to Expansion. However, in the RLR dataset, the trend from Factor to Question is in contrast with the others. A possible explanation is that the Factor description is already concrete enough and converting it into question format introduces undesired noises. Overall, the expressiveness of different forms of extraction fact labels get enhanced all the way from Token to Expansion. This consequently achieves higher scores.

5.3   Strategies of Query Expansion

Following the results, we further discuss different types of query expansion and the corresponding detailed strategies. Table 4 proves that data augmentation-based approaches outperform those synonym-based. Unexpectedly, synonym-based approaches sometimes impair the performance. Different from synonym conversion, the augmentation adds extra meaningful words to the questions. For instance in the RLR dataset, the question "劳动者的基本工资数额是多少?" (What is the basic salary of workers?) turns out to be "劳动者工资的基本数额是多少?", "劳动者的基本工资是多少?" and "劳动者每月工资基数是多少?", where $K = 3$. Although the subject of the sentence, "劳动者" (workers), does not change to its synonyms, yet the object "基本工资数额" (the amount of basic salary) becomes "工资的基本数额" (the basic amount of salary), "基本工资" (basic wage) and "工资基数" (wage base). And in the third enhanced sentence, a new word "每月" (monthly) is added as the attributive of "工资基数" (wage base). Obviously, data augmentation brings richer semantic features to the original questions, including paraphrasing of the original words and introducing of extra information. These make the questions more accurate and specific.

Nevertheless, the more is not always the better. We discover that different augmentations have inconsistent effects on different datasets and different types of questions. Expanding the question for all the answer types (span, YES, NO and NG) does not always give the best results (see results of QE_{da}-$topK$ in

Table 4 and Appendix Table 6). In the ablation experiments, some questions are preserved and the others are expanded. We first expand all the questions to find out what types of questions end in poor results, and leave these questions not expanded while expand the rest questions who have outstanding performance. The final strategy is QE_{da}-mix, whose experimental results on all three datasets get improved by up to 4 points in both F1 and EM.

For different datasets, the same type of questions does not necessarily show similar results before and after query expansion. For example, in the RLR and REJR datasets, query expansion works very well on questions whose answers are spans, and poorly on questions whose answers are "YES", "NO" and "NG". But in the II dataset, the results are exactly the opposite. Practically, it is difficult for us to fully judge the results of the experiments through pre-speculation. For instance, we intuitively thought that it is unnecessary to expand yes/no questions because their answers are simple and naive, and the question format is uniform. However, as far as the experimental results are concerned, our speculation can only be incompletely wrong. The performance of different types of questions before and after query expansion could be inconsistent, and cannot be predicted despite of the datasets they belong to.

6   Conclusion

In this paper, we design a unified framework to extract different types of legal factors. It is a MRC-based framework but with pre-determined extraction fact labels according to the datasets and achieves automatic extraction without manually feeding the questions. To verify the effectiveness of our approach, we constructed three datasets, including intentional injury (II), recourse for labor remuneration (RLR), and refusal to execute judgments or rulings (REJR) cases. Experiments show that such model suits well to our information extraction task compared to the sequence labeling models. To improve the model performance, we design different strategies and conduct plentiful experiments that discuss the impact of adding semantic information to the extraction fact labels. The strategies fall into two main categories, synonym-based and data augmentation-based. Among all the strategies of these two categories, the QE_{da}-mix performs generally well, improving the performance of up to 4% on each dataset.

7   Future Works

Our proposed framework aim at extracting different types of legal factors. So far, it is able to extract span-like factors, deduce yes/no factors and judge whether a factor is included in the document or not. However, there are still several types of legal factors that are not supported, such as multiple-choice factors, numerical derivation factors, summarisation factors, etc. multiple-choice factors refer to the fact that involves a fixed number of parties. For instance, some disputes over rental contracts often involve three parties: the landlord, the tenant, and the intermediary. The numerical derivation factors usually ask for the total amount of money, weight, and the number of criminal activities. These information are not directly and clearly written, nor can they be derived during straightforward extractions. In the future, we will manage to expand the framework's capability to let it handle more types of factors.

Experiments show that although query expansion has a positive effect on each dataset, yet different datasets have different responses to query expansion. For the RLR and REJR datasets, query expansion works well on questions whose answers are spans. While query expansion gives a good result on the other types of instance for II dataset. We believe this is due to differences in the structures and content of different case types, but we have not quantified these differences or looked for common patterns. In the future, we will try to find a universal query expansion strategy to streamline the current solution, as well as adding new factor types into consideration.

Acknowledgments. The authors would like to thank all the reviewers for their insight reviews. This paper is funded by National Key R&D Program of China (No.2018YFC0807701).

Appendix

Table 5. Data augmentation of other two datasets. We selected two representative questions from each dataset. We do not translate sentences after data augmentation (DA) because DA is usually a paraphrasing of the original questions, and the English version could remain unchanged.

Cases	Original Question	Data Augmentation(K=5)
II	被告人作案工具是什么? (What is the defendant's tool of committing the crime?)	什么是被告人的作案工具?
		被告人作案的工具都有什么?
		被告人作案工具?
		被告人作案时应该用什么工具?
		被告人作案工具有哪些?
	损失的经济构成? (What is the economic composition of the loss?)	损失的经济构成?
		损失经济构成要点?
		损坏经济损失的构成要件?
		经济损失的构成要件有哪些?
		经济损失的构成要素是什么?
REJR	被告人义务内容? (What are the obligations of the defendant?)	被告人义务的内容?
		被告人的义务是?
		被诉人义务的内容包括哪些?
		被告人的义务有哪些?
		被告人的义务是什么?
	是否对被告人限制高消费? (Is the defendant restricted to high consumption?)	被告人限制高消费是否有效?
		被告人是否能够限制高消费?
		法院对被告人高消费,是否有限制的?
		对被告人的高消费行为是否有规定?
		对被告人限制高消费的行为有哪些?

Table 6. Results of general questions, reported in **Exact Match**(%), **F1** score(%), **Precision**(%) and **R**ecall(%). The meaning of each experiment in this table is the same as that in Table 4. It is noted that the extraction results of the questions whose answer is no in the datasets of recourse for labor remuneration and refusal to execute judgments or rulings are poor, and even the results of many experiments are 0. As shown in Table 2, the questions with negative answer are originally very rare, and we do not build the datasets specifically for this situation, which made it difficult to extract, or simply do not show in the test sets.

Cases	Methods	Yes			No		
		R	P	F1	R	P	F1
Recourse for Labor Remuneration (RLR)	Q_{ori}	97.96	92.31	95.05	0.00	0.00	0.00
	$Q_{ori}+F$	100.00	90.74	95.15	0.00	0.00	0.00
	Q_{ori}-$modify$	100.00	94.23	**97.03**	16.67	100.00	28.57
	QE_{syn}-M_1	100.00	83.05	90.74	0.00	0.00	0.00
	QE_{syn}-M_2	97.96	94.12	96.00	33.33	50.00	**40.00**
	QE_{syn}-M_2+F	100.00	87.50	93.33	0.00	0.00	0.00
	QE_{da}-$random$	100.00	89.09	94.23	0.00	0.00	0.00
	QE_{da}-$top5$	100.00	90.74	95.15	0.00	0.00	0.00
	QE_{da}-$lstm$	100.00	92.45	96.08	0.00	0.00	0.00
	QE_{da}-gru	100.00	92.45	96.08	0.00	0.00	0.00
	QE_{da}-mix	100.00	92.45	96.08	0.00	0.00	0.00
	QE_{da}-yn	100.00	92.45	96.08	0.00	0.00	0.00
	QE_{da}-$top3$	97.96	92.31	95.05	0.00	0.00	0.00
	QE_{da}-$top1$	100.00	90.74	95.15	0.00	0.00	0.00
	QE_{da}-$last5$	100.00	92.45	96.08	0.00	0.00	0.00
Intentional Injury (II)	Q_{ori}	100.00	90.00	94.74	100.00	100.00	100.00
	Q_{ori}-$modify$	100.00	90.00	94.74	100.00	100.00	100.00
	QE_{da}-$top5$	100.00	90.00	94.74	100.00	100.00	100.00
	QE_{da}-$lstm$	100.00	90.00	94.74	100.00	100.00	100.00
	QE_{da}-gru	100.00	100.00	**100.00**	100.00	100.00	100.00
	QE_{da}-mix	100.00	90.00	94.74	100.00	100.00	100.00
Refusal to Execute Judgments or Rulings (REJR)	Q_{ori}	95.45	76.36	84.84	0.00	0.00	0.00
	Q_{ori}-$modify$	95.45	77.78	**85.71**	0.00	0.00	0.00
	QE_{da}-$top5$	95.45	77.78	**85.71**	0.00	0.00	0.00
	QE_{da}-$lstm$	90.91	80.00	85.11	0.00	0.00	0.00
	QE_{da}-gru	95.45	77.78	**85.71**	0.00	0.00	0.00
	QE_{da}-mix	95.45	77.78	**85.71**	0.00	0.00	0.00

References

1. Buck, C., et al.: Ask the right questions: active question reformulation with reinforcement learning. In: 6th International Conference on Learning Representations, ICLR 2018, Vancouver, BC, Canada, 30 April–3 May 2018, Conference Track Proceedings. OpenReview.net (2018). https://openreview.net/forum?id=S1CChZ-CZ

2. Che, W., et al.: "N-LTP: a open-source neural Chinese language technology platform with pretrained models. arXiv preprint arXiv:2009.11616 (2020)
3. Devlin, J., et al.: BERT: pre-training of deep bidirectional transformers for language understanding. In: Proceedings of the 2019 Conference of the North American Chapter of the Association for Computational Linguistics: Human Language Technologies, (Long and Short Papers), Minneapolis, Minnesota, vol. 1, pp. 4171–4186. Association for Computational Linguistics (2019). https://doi.org/10.18653/v1/N19-1423. https://www.aclweb.org/anthology/N19-1423
4. Duan, X., et al.: CJRC: a reliable human-annotated benchmark DataSet for Chinese judicial reading comprehension. In: CCL (2019)
5. Gao, J., Nie, J.-Y.: Towards concept-based translation models using search logs for query expansion (2012). https://www.microsoft.com/en-us/research/publication/towards-concept-based-translation-models-using-search-logs-query-expansion/
6. Grishman, R.: Information extraction: techniques and challenges. In: Pazienza, M.T. (ed.) SCIE 1997. LNCS, vol. 1299, pp. 10–27. Springer, Heidelberg (1997). https://doi.org/10.1007/3-540-63438-X_2 ISBN 978-3-540-69548-6
7. Hermann, K.M., et al.: Teaching machines to read and comprehend. In: Proceedings of the 28th International Conference on Neural Information Processing Systems, NIPS 2015, Montreal, Canada, vol. 1, pp. 1693–1701. MIT Press (2015)
8. Levow, G.-A.: The third international Chinese language processing bakeoff: word segmentation and named entity recognition. In: Proceedings of the Fifth SIGHAN Workshop on Chinese Language Processing, Sydney, Australia, July 2006, pp. 108–117. Association for Computational Linguistics (2006). https://www.aclweb.org/anthology/W06-0115
9. Li, X., et al.: FLAT: Chinese NER using flat-lattice transformer. In: Proceedings of the 58th Annual Meeting of the Association for Computational Linguistics, July 2020, pp. 6836–6842. Association for Computational Linguistics (2020). https://doi.org/10.18653/v1/2020.acl-main.611. https://www.aclweb.org/anthology/2020.acl-main.611
10. Li, X., et al.: A unified MRC framework for named entity recognition. In: Proceedings of the 58th Annual Meeting of the Association for Computational Linguistics, July 2020, pp. 5849–5859. Association for Computational Linguistics (2020). https://doi.org/10.18653/v1/2020.acl-main.519. https://www.aclweb.org/anthology/2020.acl-main.519
11. Liu, B., et al.: A user-centered concept mining system for query and document understanding at Tencent. In: Teredesai, A., et al. (eds.) Proceedings of the 25th ACM SIGKDD International Conference on Knowledge Discovery & Data Mining, KDD 2019, Anchorage, AK, USA, 4–8 August 2019, pp. 1831–1841. ACM (2019). https://doi.org/10.1145/3292500.3330727
12. Rajpurkar, P., Jia, R., Liang, P.: Know what you don't know: unanswerable questions for SQuAD. In: Proceedings of the 56th Annual Meeting of the Association for Computational Linguistics (Volume 2: Short Papers), Melbourne, Australia, July 2018, pp. 784–789. Association for Computational Linguistics (2018). https://doi.org/10.18653/v1/P18-2124. https://www.aclweb.org/anthology/P18-2124
13. Rajpurkar, P., et al.: SQuAD: 100,000+ questions for machine comprehension of text. In: Proceedings of the 2016 Conference on Empirical Methods in Natural Language Processing. Austin, Texas, November 2016, pp. 2383–2392. Association for Computational Linguistics (2016). https://doi.org/10.18653/v1/D16-1264. https://www.aclweb.org/anthology/D16-1264
14. Su, J.: SimBERT: integrating retrieval and generation into BERT. Technical report (2020)

15. Vaswani, A., et al.: Attention is all you need. In: Guyon, I., et al. (eds.) Advances in Neural Information Processing Systems 30: Annual Conference on Neural Information Processing Systems 2017, Long Beach, CA, USA, 4–9 December 2017, pp. 5998–6008 (2017). https://proceedings.neurips.cc/paper/2017/hash/3f5ee243547dee91fbd05%203c1c4a845aa-Abstract.html

16. Wu, Y., et al.: Chinese named entity recognition based on multiple features. In: Proceedings of the Conference on Human Language Technology and Empirical Methods in Natural Language Processing, HLT 2005, Vancouver, British Columbia, Canada, pp. 427–434. Association for Computational Linguistics (2005). https://doi.org/10.3115/1220575.1220629

17. Yan, H., et al.: TENER: adapting transformer encoder for named entity recognition (2019). arXiv:1911.04474 [cs.CL]

18. Zhang, Y., Yang, J.: Chinese NER using lattice LSTM. In: Proceedings of the 56th Annual Meeting of the Association for Computational Linguistics (Volume 1: Long Papers), Melbourne, Australia, July 2018, pp. 1554–1564. Association for Computational Linguistics (2018). https://doi.org/10.18653/v1/P18-1144. https://www.aclweb.org/anthology/P18-1144

19. 田久乐 and 赵蔚. "基于同义词词林的词语相似度计算方法". In: 吉林大学学报.信息科学版, vol. 06, pp. 60–66 (2010)

Meta-learned ID Embeddings for Online Inductive Recommendation

Jingyu Peng, Le Wu$^{(\boxtimes)}$, Peijie Sun, and Meng Wang

School of Computer Science and Information Engineering,
Hefei University of Technology, Hefei 230009, China

Abstract. Learning accurate user and item ID embeddings from user-item historical records has shown great success for recommender systems. Most of these embedding learning models are transductive and work well for users that appear in the training stage. However, in the model serving stage, new users continue to join the system. It's important to quickly adapt to new users' preferences for online inductive recommendation scenarios. Some previous works adopted embedding retraining or fed content data for new user ID embedding learning, these models either suffered from slow convergence or relied on auxiliary data. In this paper, we propose a meta-learned ID embedding framework for new users without using any side information in online inductive recommendation scenarios. Our key idea is that, we treat each user's ID embedding learning as a separate task, and propose to meta-learn the initial embedding by modeling the global knowledge from all users (tasks). Each user's embedding is initialized by the learned global knowledge instead of randomly initialization. Therefore, we could quickly adapt to a new user's ID embedding based on a few updates from her online records, which can facilitate fast online recommendation. Moreover, our main technical contribution lies in how to learn the global prior knowledge for informative ID embedding initialization without any side information. Finally, extensive experimental results on three real-world datasets clearly show both the efficiency and effectiveness of the meta-learned ID embeddings for inductive recommendation.

Keywords: Recommender system · Meta learning · Collaborative Filtering · Cold-start

1 Introduction

Collaborative Filtering (CF) is a popular approach for building recommender systems, with the assumption that users' preferences for items could be collaboratively modeled from users' historical behavior data [1,20]. Among all CF models, learning accurate user and item ID embeddings has been the key technology that dominates CF area [3,15,18,24,25]. These embedding models can learn low dimensional dense vector representations of users from their past behavior. However, most of these ID embedding based models are naturally transductive,

© Springer Nature Switzerland AG 2021
H. Lin et al. (Eds.): CCIR 2021, LNCS 13026, pp. 32–44, 2021.
https://doi.org/10.1007/978-3-030-88189-4_3

meaning that during the model serving process, each test user must appear in the training process. In practical recommender systems, new users continuously join the platform, e.g., a new user registers to or an anonymous user enters a platform, and shows preferences to some items (e.g., browses or buys several items). It is very cruical to update recommender systems timely to serve these new users, as it can improve user satisfaction and increase their loyalty to the platform.

To achieve the inductive learning with new users at test time, some researchers provided content-based approaches to learn new users' preferences, but most new or anonymous users are reluctant to fill any personal information [6,22]. Another naive idea is to retrain recommender models with new users' behavior. As full retraining is time consuming, an alternative solution is only to update new users' embeddings while keeping the embeddings of items and old users fixed. For each new user, the training process usually starts with random initial user embedding, and the model performance needs many update times to reach a local minimum. Although this fine-tune process shows comparable efficiency compared to the full data retraining, obtaining new users' final ID embedding with many training epochs is still far from the online latency requirements. In summary, relying on auxiliary data or suffering from the time efficiency issue make current models inferior choices for serving new users online.

In this paper, we explore whether it is possible to provide fast recommendation for new users in inductive recommendation without using any side information. Instead of randomly initializing ID embeddings for new users, our high-level idea is to learn better initial embeddings for new users, such that to speed up the learning process of new users with very limited records. As predicting each user's preference to an item can be regarded as a classification problem, we treat each user's embedding learning as a separate task, and make an analogy between recommending some products to a new user with few interactions and few-shot classification [5,12]. Therefore, it is natural to apply meta-optimization approaches, which are successful in fast adaption of few-shot classification [5]. The core idea of meta-optimization approaches is to train global sharing initialization parameters for all tasks (users). When a new user (task) comes, her ID embedding could be initialized with the global learned knowledge, then her final ID embedding can be quickly to be adapted with a few updates to facilitate fast online recommendation.

With the analogy between recommending some products to a new user with few interactions and few-shot classification, there are several recent attempts that leveraging meta-optimization approaches to generate better initial embeddings for new users, either with the entity profile information [11,17] or with auxiliary heterogeneous information networks [13]. Nevertheless, it is non-trivial to apply these meta-optimization techniques, as we do not have any content input for new users. How to define the general sharing initialization parameters for all users becomes the key challenge. To tackle this challenge, we design two detailed strategies for global parameters initialization. The first model is straightforward by treating the initialization ID embedding as global parameter for all users, i.e., all users share the same initial ID embedding. The second idea is feeding pretrained item embeddings and the current user's limited records to learn the global parameters that can be used to output the unique ID embedding of each user. Finally, we conduct extensive experiments on three real-world datasets and the experimental results clearly show the effectiveness and efficiency of our proposed framework. For example, our proposed framework could improve the recommendation accuracy with more than 10% and the training efficiency with less than one-tenth time cost compared to the best baseline.

2   Preliminaries

Given the user set $\mathcal{U} = (1, 2, 3, ..., M), |\mathcal{U}| = M$ and item set $\mathcal{V} = (1, 2, 3, ..., N), |\mathcal{V}| = N$, their interaction records form the rating matrix $\mathbb{R}^{M \times N}$. In this matrix, r_{uv} equals one when user u rates item v but zero otherwise. For each user u, we use $\mathcal{R}_u^+$ to denote the positive itemset that u shows preferences, i.e., $\forall v \in V, v \in \mathcal{R}_u \iff r_{uv} = 1$. And we randomly select k times the size of $\mathcal{R}_u^+$ items from the set $\mathcal{V} - \mathcal{R}_u^+$. The selected items are treated as the negative itemset $\mathcal{R}_u^-$.

With users' preferences, state-of-the-art CF embedding models focus on learning a low dimensional embedding space of users and items, namely $\mathbf{P} \in \mathbb{R}^{D \times M}$ and $\mathbf{Q} \in \mathbb{R}^{D \times N}$. Then, the predicted preference of the user-item pair (u, v) is modeled by the inner product of their corresponding embedding vectors as:

$$\hat{r}_{uv} = \mathbf{p}_u^T \mathbf{q}_v. \tag{1}$$

The widely used binary cross-entropy loss [7] is adopted as the loss function as:

$$\mathcal{L} = -\sum_{u=1}^{M} \sum_{i \in \mathcal{R}_u^+} \sum_{j \in \mathcal{R}_u^-} (r_{ui} \log(\hat{r}_{ui}) + (1 - r_{uj}) \log(1 - \hat{r}_{uj})) + \lambda \|\mathbf{P}\|_F^2 + \lambda \|\mathbf{Q}\|_F^2, \tag{2}$$

where the first term captures the training loss, the last two terms are l2-norm regularization terms with model parameters as $[\mathbf{P}, \mathbf{Q}]$, and λ is a regularization parameter.

These CF models perform well for the transductive setting, i.e., all test users appear in the training data. However, in the real world, inductive learning is more general with new users continuously join the system and show their preferences with very limited records in a short session. How to provide timely recommendations for new users has become a critical issue. A straightforward solution is to retrain the embedding-based recommendation model with both the online new user data and offline data. However, the retraining time is time-consuming. An alternative solution is to only learn the ID embeddings of new users and keep learned item embeddings fixed as the item embeddings have been well trained offline. Let a denote a new user that does not appear offline, i.e., $a \notin \mathcal{U}$, our goal is to learn the new user ID embedding $\mathbf{p}_a$ by minimizing the following function:

$$\mathcal{L}_a = -\sum_{i \in \mathcal{R}_a^+} \sum_{j \in \mathcal{R}_a^-} (r_{ai} \log(\hat{r}_{ai}) + (1 - r_{aj}) \log(1 - \hat{r}_{aj})) + \lambda \|\mathbf{p}_a\|_F^2, \tag{3}$$

where $\mathbf{p}_a$ is the embedding of new user a that does not appear in the training data, and λ is the same regularization parameter as Eq. (2).

Please note that, as each new user has very limited records, the average number of most new users' rating records is far less than the ID embedding size D, i.e., $|\mathcal{R}_a^+| \ll D$. Given the optimization function for each new user a, we start with a random initialization of $\mathbf{p}_a$, and perform gradient descent until convergence. The training process will cost hundred of update epochs. Thus, it could not satisfy users' real-time needs for serving new users online. It is natural to ask the question: could we design a fast learning model for new users, such that we could quickly learn a new user's preference with a few gradient steps?

3   Meta-learned ID Embeddings for New Users

In this section, we first propose how to recast the problem of fast inductive recommendation with new users under the meta-learning framework. Then, we give two detailed architectures of designing the meta-learned ID embedding models without any content input. After that, we briefly show how to quickly adapt to new users' ID embeddings at online serving stage with the learned meta-knowledge.

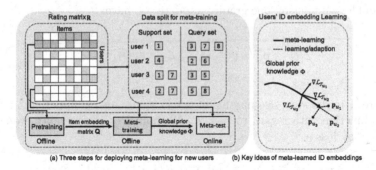

Fig. 1. The overall framework of our proposed framework, with the key idea of meta-learned ID embedding is shown at the right part of this figure.

3.1   Meta-learning for Inductive Recommendation

To quickly learn each new user's embedding vector with her limited rating records, we could build a connection between meta-learning and inductive recommendation for new users. By treating each user u's ID embedding learning as a task, each task (user) has very limited training data $\mathcal{R}_u^+$. Meta-learning provides a potential solution to our problem: by learning to learn across data from many previous tasks (users), meta-learning algorithms can discover the global meta-knowledge among tasks (all training users) to enable fast learning on new tasks (new users). In our framework, the prior knowledge is denoted as a parameter set Φ, and instead of random initialization of each user's embedding $\mathbf{p}_u^0$ as previous works, we learn a function g parameterized by Φ to initialize ID embedding vector $\mathbf{p}_u^0 = g(; \Phi)$.

For ease of clarification, in Fig. 1, we show the concrete steps for deploying meta-learning for new users. There are three steps: pretraining, meta-training and meta-test. These three steps follow a natural time line. In the pretraining step, we can adopt any embedding models to learn user and item embeddings. This step outputs item embedding matrix $\mathbf{Q}$ for the following two steps. Next, we mimic the meta-training process to divide the data of each task into a support set and a query set, and design meta-learning framework to learn the global knowledge. The global knowledge is then sent to the online stage to initialize the new user's ID embedding, which can facilitate quickly adapting to new users' ID embedding learning through one or more gradient steps.

Specifically, during meta training, similar to the setting in MAML, we split the original training data of each task $\mathcal{T}_u$ into two sets: a support set $\mathcal{S}_u$ and a query set $\mathcal{Q}_u$. The support set and query set come from u's rated itemset $\mathcal{R}_u^+$ and are mutually exclusive: $\mathcal{S}_u \cap \mathcal{Q}_u = \emptyset, \mathcal{S}_u, \mathcal{Q}_u \subseteq \mathcal{R}_u^+$. Each task $\mathcal{T}_u$ is associated with task-specific local parameters, i.e., the ID embeddings $\mathbf{p}_u$. We use $\mathbf{p}_u^0$ to denote the initial embedding of the local parameter $\mathbf{p}_u$. Then, this task updates its local parameter $\mathbf{p}_u$ with the support set $\mathcal{S}_u$ using one or a few gradient steps. For example, when using one gradient update, we have:

$$\begin{aligned} \mathbf{p}_u' &= \mathbf{p}_u - \alpha \nabla_{\mathbf{p}_u} \mathcal{L}_{\mathcal{T}_u}(g(; \Phi)) \\ &= \mathbf{p}_u - \alpha \nabla_{\mathbf{p}_u} \mathcal{L}_{\mathcal{T}_u}(\mathbf{p}_u^0), \end{aligned} \tag{4}$$

where α is the step size parameter for local parameter update, and $\mathcal{L}_{\mathcal{T}_u}$ is the loss function with regard to task $\mathcal{T}_u$. Without loss of generality, we use the cross-entropy loss as:

$$\mathcal{L}_{\mathcal{T}_u} = - \sum_{i \in \mathcal{S}_u} \sum_{j \in \mathcal{S}_u^-} (r_{ui} log(\hat{r}_{ui}) + (1 - r_{uj}) log(1 - \hat{r}_{uj})), \tag{5}$$

where $\mathcal{S}_u^-$ is k times of the size of $\mathcal{S}_u$, and $\mathcal{S}_u^- \subseteq \mathcal{R}_u^-$. In the above equation, we do not have any regularization term as Eq. (3). The reason is that, meta-learning algorithms only perform several gradient steps based on Eq. (5), and works as a early stopping without any overfitting issue.

For each task $\mathcal{T}_u$, after learning the updated local parameter $\mathbf{p}_u'$, we learn how to learn the performance of global parameters Φ with the query set $\mathcal{Q}_u$. The global parameters Φ is trained by optimizing the performance of updated local parameters $\mathbf{p}_u'$ with respect to Φ across all tasks (users).

$$\arg \min_{\Phi} \sum_{u \in \mathcal{Q}_u} \mathcal{L}_{\mathcal{T}_u}(\mathbf{p}_u') = \mathcal{L}_{\mathcal{T}_u}(\mathbf{p}_u - \alpha \nabla_{\mathbf{p}_u} \mathcal{L}_{\mathcal{T}_u}(g(; \Phi))). \tag{6}$$

In the above equation, please note that the meta-optimization performance is evaluated on the updated task-specific local parameters $\mathbf{p}'_u$, which are learned from current global parameters Φ (e.g., a step update with Eq. (4)). Then, meta-optimization over tasks (users) is also updated with stochastic gradient descent as:

$$\Phi \leftarrow \Phi - \beta \nabla_\Phi \sum_{u \in \mathcal{Q}_u} \mathcal{L}_{\mathcal{T}_u}(\mathbf{p}'_u), \tag{7}$$

where β is the meta step size.

3.2   Architecture of Meta-Learned ID Embeddings

Given the formulation above, the problem of meta-learned ID embedding framework turns to how to build a function $g(; \Phi)$ to extract global knowledge structure for initial user ID embedding of each task $\mathcal{T}_u$ as: $\mathbf{p}^0_u = g(; \Phi)$.

General Learner. Without any side information as input, a simple idea of the global knowledge learner is to set g to an identity function. Formally, for any task $\mathcal{T}_u$ of a user u, we have:

$$\mathbf{p}^0_u = g_1(; \Phi) = \mathbf{I}\Phi = \Phi, \tag{8}$$

where $\mathbf{I} \in \mathbb{R}^{D \times D}$ is an identity matrix, and $\Phi \in \mathbb{R}^D$. In other words, we assume that there exists similarities of all users, and it is presented in the form that each user has a same preference initialization vector.

Personalized Learner. The general learner is simple, but its expressiveness may be limited by assigning the same initialization vector for all users. As each user u has limited available records $\mathcal{S}_u$, we design a personalized embedding learner to fully utilizing her rating records. Since we already pretrained item embedding matrix $\mathbf{Q}$, the personalized initialization vector for each user can be calculated with:

$$\mathbf{x}_u = Pooling(\mathbf{Q}[\mathcal{S}_u]) \tag{9}$$
$$\mathbf{p}^0_u = g_2(\mathbf{x}_u; \Phi), \tag{10}$$

where $\mathbf{Q}[\mathcal{S}_u]$ denotes the sub item embedding matrix As each user's support set varies, the pooling operation in Eq. (9) transforms the variable length submatrix into a fixed size vector output. Both the pooling function and the learner g_2 can be flexible. In practice, we choose average pooling as it achieves better performance compared to max pooling. Equation (10) could be a linear function as $\mathbf{p}^0_u = \mathbf{W}\mathbf{x}_u$ with transformation matrix $\mathbf{W}$, or a multilayer perceptron to capture the non-linear relationships. Compared to the general learner, the personalized learner utilizes more personalized information for initial embedding learning.

Algorithm 1. Training Process of Meta-Learned ID Embedding

Input: Task $\mathcal{T}_u$ with support and query set

Input: Pretrained item-embedding matrix $\mathbf{Q}$

Input: Step size hyperparameters α and β

Input: The local update times K

Output: The shared global parameters Φ

1: Randomly initialize global parameter Φ

2: **while** Not converge **do**

3:    Randomly sample batch of users $\mathcal{B} \subset \mathcal{U}$

4:    **for** user u in $\mathcal{B}$: **do**

5:        Initialize $\mathbf{p}_u^0 = g(; \Phi)$ based on a detailed architecture;

6:        $\mathcal{L} = 0$;

7:        **for** $k = 1; k \leq K; k++$ **do**

8:            **for** $v \in V$: **do**

9:                $\hat{r}_{uv} = \mathbf{q}_V^T \mathbf{p}_u^{k-1}$;

10:           **end for**

11:           Calculate loss $\mathcal{L}_u$ based on Eq.(5);

12:           $\mathbf{p}_u^k \leftarrow \mathbf{p}_u^{k-1} - \alpha \nabla_{\mathbf{p}_u^{k-1}} \mathcal{L}_{\mathcal{T}_u}$;

13:       **end for**

14:       $\mathbf{p}_u = \mathbf{p}_u^K$;

15:       **for** $v \in V$: **do**

16:           $\hat{r}_{uv} = \mathbf{q}_V^T \mathbf{p}_u$;

17:       **end for**

18:       Calculate $\mathcal{L}_u$ with Eq.(5) based on query set $\mathcal{Q}_u$;

19:       $\mathcal{L} = \mathcal{L} + \mathcal{L}_{\mathcal{T}_u}$

20:    **end for**

21:    $\Phi \leftarrow \Phi - \beta \nabla_\Phi \mathcal{L}$;

22: **end while**

23: **Return** Global parameter set Φ.

We show the details of meta-training in Algorithm 1. For the above two detailed architectures, the only difference in this algorithm is calculating $g(; \Phi)$ in Line 5. In practice, for each user u, the unobserved feedbacks $\mathcal{V} - \mathcal{S}_u$ is much larger than the observed support set $\mathcal{S}_u$ in Eq. (5). Similar as many previous works [3,7,19], we randomly select 3 times of the size of $\mathcal{S}_u$ as possible negative items at each training epoch.

3.3   Meta-test Stage

After finishing the meta-training process, we get the global parameter set Φ. For online serving stage, if a new user a comes and shows preferences to a limited item set $\mathcal{S}_a$, we could initialize her embedding as $\mathbf{p}_a^0 = g(; \Phi)$ and quickly update her embedding with K gradient steps.

4   Experiments

4.1   Experimental Settings

Datasets. We conduct experiments on three real-world datasets: *MovieLens-1M*[1], *Amazon Cell Phones and Accessories* and *Amazon CDs and Vinyl*[2]. In the following subsections, the MovieLens-1M, Amazon Cell Phones and Accessories, and Amazon CDs and Vinyl are called MovieLens, Amazon Small and

---

[1] https://grouplens.org/datasets/movielens/1m/.

[2] http://jmcauley.ucsd.edu/data/amazon/.

Table 1. The statistics of the three datasets.

Datasets		MovieLens	Amazon small	Amazon Big
Pre-training	Users	1,510	6,970	18,815
	Items	3,952	9,448	57,750
	Ratings	249,088	59,606	252,706
Meta-training	User	3,020	13,937	37,626
	Ratings	18,316	59,606	182,832
	Avg size of support set	2.81	1.93	2.21
	Avg size of query set	3.25	2.34	2.65
Meta-test	Users	1,510	6,953	18,749
	Ratings	8,978	33,497	97,640
	Avg size of support set	2.70	2.15	2.35
	Avg size of query Set	3.20	2.67	2.85

Amazon Big for short. As we focus on the ranking task, for all datasets, we transform the ratings in the original datasets into implicit feedback. If one user rates an item, the corresponding entry will be treated as 1, otherwise it will be 0. As illustrated in Fig. 1, there are three steps in deploying meta-learning for new users: pretraining, meta-training, and meta-test. Since there should be no duplicate users among these three steps, we randomly split all users into three parts in the ratio 1:2:1 for pretraining, meta-training, and meta-test, respectively. In the pretraining stage, we randomly select one record of each user as the validation data. In the meta-training stage, for each user we randomly select 2 to 10 historical records of her. Half of the selected records are treated as the support set, and the rest of the selected records are treated as the query set. And we adopt the same procedure to prepare the support set and query set for each user in the meta-test stage. Details of all dataset are shown in Table 1.

Experimental Setup. We call our proposed framework of meta-learned ID embeddings MetaCF for inductive CF. The two detailed architectures for MetaCF are denoted as MetaCF_G for general embedding (Eq. (8)) and MetaCF_P for personalized embedding (Eq. (10)). As (10) could be a linear function or a multilayer perceptron to capture the non-linear relationships, we use MetaCF_P(Linear) and MetaCF_P(Neural) to denote these two choices. The neural architecture is a two-layered neural network with ReLU activation. To study the efficiency and effectiveness of our proposed model, the classical recommender model Bayesian Personalized Ranking [19] and the meta-learning based model MeLU [11] are chosen as the baseline models. For fair comparison, the two baselines use the same pretrained item embedding matrix as our proposed framework. They start with random initialization of new users with meta-test data, and use the same prediction function as our proposed framework with the similar loss function (Eq. (3)). The performance of all models is evaluated on the query set of the users in the meta-test stage. The meta-learning based models, i.e., our proposed model and MeLU, are trained on the meta-train data first.

Then for each new user in the meta-test stage, the support set of each user is used to learn her ID embedding. However, BPR is only trained on the support set of all users in the meta-test stage to learn the ID embeddings of all users. Hit Ratio (HR) and Normalized Discounted Cumulative Gain (NDCG) [26,27] are used to evaluate the performance of all models. For both metrics, a larger value means a better performance.

Parameter Setting. For all models, we set the dimension D to 32. The step size alpha is set to 2×10^{-3} for all models and beta is set to 1×10^{-7}, 1×10^{-6}, and 1×10^{-5} for MetaCF_G, MetaCF_P(Linear), and MetaCF_P(Neural), respectively. The local updates times K varies from one to five. Please note that, as the meta-gradient involves second derivatives when performing back gradient over the meta-objective (Eq. (5)), we resort to first-order approximation, which shows nearly the same performance as obtained with full second derivatives [5].

Table 2. Overall performance of our proposed models. Bold font means the best model and underline means the corresponding model ranks second.

Models	MovieLens		Amazon small		Amazon Big	
	HR@10	NDCG@10	HR@10	NDCG@10	HR@10	NDCG@10
BPR_5	0.0341	0.0258	0.0195	0.0130	0.0047	0.0031
BPR_Best	<u>0.0656</u>	0.0462	<u>0.0308</u>	**0.0209**	0.0070	0.0045
MeLU	0.0220	0.0108	0.0048	0.0029	0.0022	0.0013
MetaCF_G	0.0444	0.0350	0.0247	0.0178	0.0066	0.0042
MetaCF_P(Linear)	0.0565	<u>0.0465</u>	**0.0312**	0.0203	**0.0092**	**0.0060**
MetaCF_P(Neural)	**0.0694**	**0.0478**	0.0298	<u>0.0205</u>	<u>0.0081</u>	<u>0.0053</u>

4.2 Model Performance

We report the overall performance comparison of our proposed framework and baselines in Table 2. To verify whether our proposed models can quickly adapt to a new user's ID embedding based on a few updates from her online records, we report the results of our proposed models when the local update times is set as 5. The local update times of MeLU is also set to 5. We report two kinds of results of the BPR model. BPR_5 denotes when the training epoch is set as 5. And BPR_Best denotes the best performance of BPR without training epoch limitation.

According to the results in Table 2, we have the following conclusions. First, compared with BPR_5 and MeLU, all our proposed models have significant improvements on three datasets under all the metrics. The epochs of the BPR_Best for MovieLens, Amazon Small, and Amazon Big are 44, 49, and 50, respectively. Although the local update times of our proposed model is set to 5, our proposed models perform better than BPR_Best on MovieLens and Amazon

Big datasets. E.g., on Amazon Big dataset, the NDCG@10 reaches 0.0060 for MetaCF_P(Linear), with more than 10% improvement compared to BPR_Best. We guess a possible reason of the recommendation performance gain is that, as MetaCF_P(Linear) can learn the global knowledge to speed up training for new users and the prior knowledge of all training users can help to alleviate the extreme sparsity of test users. The reason why MeLU performs worse may be that MeLU is designed for the content-based recommendation without considering any collaborative information.

When comparing the performance of our proposed three architectures, we observe that MetaCF_G with the same meta-learned initialization of all users could already reach quite good results. MetaCF_P(Linear) and MetaCF_P(Neural) perform better than MetaCF_G on MovieLens and Amazon Small datasets. By comparing the performance of MetaCF_P(Linear) and MetaCF_P(Neural), we can find the neural implementations achieve better result on Movielens and Amazon Small, while the linear implementations perform better on Amazon Big. In despite of the stronger express ability of neural architecture, the data sparsity and data size limit the performance.

Table 3. Performance with different local update times K under metrics HR@10.

Model		Local update times K				
		1	2	3	4	5
MovieLens	BPR_K	0.0110	0.0168	0.0236	0.0298	0.0341
	MeLU	0.0213	0.0224	0.0235	0.0219	0.0220
	MetaCF_G	0.0264	0.0341	0.0395	0.0429	0.0444
	MetaCF_P(Linear)	0.0543	0.0546	0.0560	0.0559	0.0565
	MetaCF_P(Neural)	0.0677	0.0682	0.0686	0.0691	0.0694

In Table 3, we show the performance of all models with different values of local update times K on Movielens. As K ranges from 1 to 5, we find the performance of BPR_K has a large improvement. By contrast, the performance of our proposed models is relatively stable and is not strongly influenced by K. We think this is caused by the meta-knowledge learned by our proposed models, such that one local update can already reach very good ID embedding. Based on this result, in practice with very high time request, we could set the local update times to 1.

In Fig. 2, we compare the performance of all models with different support set size on Amazon_Big. From the result, we can find the performance of all models improves stable with increasing the support set size. And under all cases, our proposed models always perform better than the baselines.

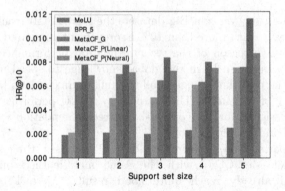

Fig. 2. Performance with different support set size under metrics HR@10 on Amazon_Big

5   Related Work

Learning low-dimensional ID embeddings of users and items from their historical behavior has been proved extremely useful for modern recommender system design [10,19]. Most CF approaches are transductive and could not apply to new nodes at test stage. To tackle the new node problem at test stage, some works are proposed to leverage node content to build a connection between content embedding and ID embedding for inductive learning [6,22]. Researchers have recently attempted to utilize sub-graph based neural network between each possible user-item pair for inductive matrix factorization with new users or items [28]. However, due to the huge time complexity of each candidate sub-graph modeling, it is impractical for online stage. To tackle the streaming data problem at model serving stage, how to incrementally update and retraining these systems is also a hot topic [8,29]. For new users at test stage, a simple idea is to keep the item embeddings learned from history fixed, while learning new user ID embedding from random initialization [9,23]. In practice, these models still cost many epochs to reach stable. Different from these works, we focus on how to quickly adapt to each new user's ID embedding with a better ID embedding initialization.

Meta-learning, based on "learning to learn" concept, learns the meta-knowledge through a variety of learning tasks [5,12,16]. Among all meta-learning approaches, Model-agnostic Meta-Learning (MAML) is an optimization based meta-learning approach that is widely used in many scenarios [5]. MAML treats the learned shared global parameter as the initial state of any task, such that the local parameters of each new task can be achieved with very few gradient steps and a small of amount of data. Meta-learning models are employed in various recommendation scenarios. Most meta-learning based approaches for recommendation focused on the cold-start recommendation with user or item entity features [2,11,17,21,30], auxiliary heterogeneous networks [13], or the sparse context data [4]. E.g., Pan et al. proposed an optimization-based method to learn an ID generator to generate desirable initial embeddings for new ad IDs based on the features of ads [17]. And MeLU is designed for content based recommendation with user preference estimator is trained with meta-learning[11]. Besides, meta-learning approaches are also used to select user-level adaptive recommendation model selection [14]. We differ greatly as we focus on fast adaption of user embeddings without any content or auxiliary data, which makes our model more general in practice.

6 Conclusions and Future Work

In this paper, we proposed a meta-learning framework for online inductive setting with new users. To the best of our knowledge, we are one of the first few attempts that provided meta-learned new user ID embedding without any content information. By recasting this problem as a meta-learning solution, we designed different architectures to transform the prior knowledge into initial ID embeddigns without any content input. Extensive experimental results on three real-world datasets clearly showed the effectiveness and efficiency of our proposed framework. In the future, we would like to design meta-learning algorithms for online inductive recommendation with both new users and new items.

References

1. Adomavicius, G., Tuzhilin, A.: Toward the next generation of recommender systems: a survey of the state-of-the-art and possible extensions. TKDE **17**(6), 734–749 (2005)
2. Bharadhwaj, H.: Meta-learning for user cold-start recommendation. In: 2019 International Joint Conference on Neural Networks (IJCNN), pp. 1–8. IEEE (2019)
3. Chen, L., Wu, L., Hong, R., Zhang, K., Wang, M.: Revisiting graph based collaborative filtering: a linear residual graph convolutional network approach. In: AAAI, vol. 34, pp. 27–34 (2020)
4. Du, Z., Wang, X., Yang, H., Zhou, J., Tang, J.: Sequential scenario-specific meta learner for online recommendation. In: SIGKDD, pp. 2895–2904 (2019)
5. Finn, C., Abbeel, P., Levine, S.: Model-agnostic meta-learning for fast adaptation of deep networks. In: ICML, pp. 1126–1135 (2017)
6. Hamilton, W., Ying, Z., Leskovec, J.: Inductive representation learning on large graphs. In: NIPS, pp. 1024–1034 (2017)
7. He, X., Liao, L., Zhang, H., Nie, L., Hu, X., Chua, T.S.: Neural collaborative filtering. In: WWW, pp. 173–182 (2017)
8. Huang, X., Wu, L., Chen, E., Zhu, H., Liu, Q., Wang, Y.: Incremental matrix factorization: a linear feature transformation perspective. In: IJCAI, pp. 1901–1908 (2017)
9. Jiang, M., Cui, P., Wang, F., Zhu, W., Yang, S.: Scalable recommendation with social contextual information. IKDE **26**(11), 2789–2802 (2014)
10. Koren, Y.: Factorization meets the neighborhood: a multifaceted collaborative filtering model. In: SIGKDD, pp. 426–434 (2008)
11. Lee, H., Im, J., Jang, S., Cho, H., Chung, S.: MeLU: meta-learned user preference estimator for cold-start recommendation. In: SIGKDD, pp. 1073–1082 (2019)

12. Li, Z., Zhou, F., Chen, F., Li, H.: Meta-SGD: learning to learn quickly for few-shot learning. arXiv preprint arXiv:1707.09835 (2017)
13. Lu, Y., Fang, Y., Shi, C.: Meta-learning on heterogeneous information networks for cold-start recommendation. In: SIGKDD, pp. 1563–1573 (2020)
14. Luo, M., et al.: MetaSelector: meta-learning for recommendation with user-level adaptive model selection. In: WWW, pp. 2507–2513 (2020)
15. Mnih, A., Salakhutdinov, R.R.: Probabilistic matrix factorization. In: NIPS, pp. 1257–1264 (2008)
16. Nichol, A., Schulman, J.: Reptile: a scalable metalearning algorithm. arXiv preprint arXiv:1803.02999 **2**(3), 4 (2018)
17. Pan, F., Li, S., Ao, X., Tang, P., He, Q.: Warm up cold-start advertisements: improving CTR predictions via learning to learn id embeddings. In: SIGIR, pp. 695–704 (2019)
18. Rendle, S.: Factorization machines. In: ICDM, pp. 995–1000 (2010)
19. Rendle, S., Freudenthaler, C., Gantner, Z., Schmidt-Thieme, L.: BPR: Bayesian personalized ranking from implicit feedback. In: UAI, pp. 452–461 (2009)
20. Sarwar, B., Karypis, G., Konstan, J., Riedl, J.: Item-based collaborative filtering recommendation algorithms. In: WWW, pp. 285–295 (2001)
21. Vartak, M., Thiagarajan, A., Miranda, C., Bratman, J., Larochelle, H.: A meta-learning perspective on cold-start recommendations for items. In: NIPS, pp. 6904–6914 (2017)
22. Volkovs, M., Yu, G., Poutanen, T.: DropoutNet: addressing cold start in recommender systems. In: NIPS, pp. 4957–4966 (2017)
23. Wang, F., Tong, H., Lin, C.Y.: Towards evolutionary nonnegative matrix factorization. In: AAAI, pp. 501–506 (2011)
24. Wang, X., He, X., Wang, M., Feng, F., Chua, T.S.: Neural graph collaborative filtering. In: SIGIR, pp. 165–174 (2019)
25. Wu, L., He, X., Wang, X., Zhang, K., Wang, M.: A survey on neural recommendation: from collaborative filtering to content and context enriched recommendation. arXiv preprint arXiv:2104.13030 (2021)
26. Wu, L., Li, J., Sun, P., Hong, R., Ge, Y., Wang, M.: DiffNet++: a neural influence and interest diffusion network for social recommendation. IEEE Trans. Knowl. Data Eng. (2020)
27. Wu, L., Sun, P., Fu, Y., Richang, H., Xiting, W., Meng, W.: A neural influence diffusion model for social recommendation. In: SIGIR, pp. 235–244 (2019)
28. Zhang, M., Chen, Y.: Inductive matrix completion based on graph neural networks. In: ICLR (2020)
29. Zhang, Y., et al.: How to retrain a recommender system? In: SIGIR, pp. 1479–1488 (2020)
30. Zhu, Y., et al.: Learning to warm up cold item embeddings for cold-start recommendation with meta scaling and shifting networks. arXiv preprint arXiv:2105.04790 (2021)

Modelling Dynamic Item Complementarity with Graph Neural Network for Recommendation

Yingwai Shiu[1], Weizhi Ma[2], Min Zhang[1]($\boxtimes$), Yiqun Liu[1], and Shaoping Ma[1]

[1] Department of Computer Science and Technology,
Institute for Artificial Intelligence,
Beijing National Research Center for Information Science and Technology,
Tsinghua University, Beijing, China
z-m@tsinghua.edu.cn
[2] Institute for AI Industry Research (AIR), Tsinghua University, Beijing, China

Abstract. Relationships among items, especially complementarity, have shown great potential to empower the performance and explainability of recommender systems. However, there are two key limitations: 1) Most previous methods use co-occurrence to quantify item complementary relationship, which lacks theoretical support and overlooks the fact that co-occurrence is only a necessary but not sufficient condition to identify item complementarity. 2) Most studies do not consider the time-sensitive nature of item complementarity, which does exist in real scenarios.

In this study, we propose a Graph Neural Network (DCGNN) to model the dynamic item complementarity for the recommendation. First, to improve the reliability of item relationships, complementary item pairs are mined according to the 'cross elasticity of demand' concept in economic theory and the mined relationships are applied to enrich the user-item graph. Second, considering the time-sensitive nature of item complementarity, we design a time-transfer mechanism to distillate historical knowledge of item complementarity by using graph neural networks. Finally, extensive experiments and analysis were conducted on two real-world data sets, which demonstrate the effectiveness of DCGNN in capturing dynamic item complementarity and recommendation.

Keywords: Recommender system · Item complementarity · GNN

1 Introduction

With the rapid development of online applications, recommender systems (RS) are deeply integrated with our daily lives. RS is a key AI application to drive user satisfaction and business intelligence as it facilitates users in finding desired items (e.g. products, services) from a huge selection of candidates easily. Besides analyzing patterns of interest in products, the understanding of inter-item relationship is getting more attention [7]. In particularly, item complementarity is useful in terms of delivering recommendations that are relevant to a specific context.

© Springer Nature Switzerland AG 2021
H. Lin et al. (Eds.): CCIR 2021, LNCS 13026, pp. 45–56, 2021.
https://doi.org/10.1007/978-3-030-88189-4_4

According to economics theory [9], a complementary item is a good/product in which its functionality is related to the use of an associated good/product. Due to the natural relatedness between items, some previous studies have attempted to identify and model complementarity in recommendation models, which shows a certain extent of effectiveness. However, we find two key limitations that hinder the practicality of existing models in real-world scenarios. Firstly, the majority of works heavily rely on co-occurrence signals to quantify the magnitude of item complementarity, which lack theoretical support. In reality, complementary relationship between items does not only mean co-purchase behaviour, but also carry a strong emphasis on co-consumption to meet a certain goal. Secondly, most previous works ignore the time-sensitive nature of item complementarity. As users' usage habits change along the time given different usage contexts, item complementarity also dynamically evolves. In other words, for any single item, the user will buy different combinations of complementary item sets in different periods. Take tennis shoes as an example, people tend to purchase different sets of complementary clothing items in summer (e.g. t-shirts, shorts) and winter (e.g. sweatshirts, windbreakers). This implies long-term historical purchase behaviour does not necessarily reflect current complementary relationships.

To tackle the aforementioned challenges, in this paper, we propose a novel graph-based Dynamic Complementary Graph Neural Network (DCGNN) to model dynamic item complementarity in recommender systems. For the first limitation, to alleviate the reliance on co-occurrence signal and acquire functional complementary item pairs, DCGNN utilizes item complementarity according to the cross elasticity demand concept from the perspective of economics theory. Specifically, we design a graph neural networks by incorporating complementary item-item edges, in which only strong complementary item pairs are retained. For the second limitation, to consider the time-sensitive nature of item complementarity, DCGNN also enables an effective time-transfer training mechanism to inherit and incrementally adjust historical knowledge of complementarity for the recommendation task. To summarize, our main contributions are as follows:

- This is the first study that addresses the importance of the time-sensitive nature of item complementarity. Besides, a graph-based DCGNN is proposed to handle dynamic item complementarity for the task of recommendation.
- We designed a novel time-transfer mechanism in DCGNN, which performs knowledge distillation by transferring historical knowledge of item complementarity to the graph neural network. It reduces the amount of training data needed for the final prediction and is also flexible to other models.
- Extensive experiments are conducted on two real-world data sets to demonstrate the effectiveness of modelling the dynamics of item complementarity.

2 Related Work

2.1 Relation-Based Neural Recommendation Models

Recommender systems (RS) is a trending research topic and a wide spectrum of models have been proposed [1]. To pursue a more explainable and better-performed RS, recent efforts start studying and exploiting the relationship between products [2,15]. The typical ones include substitutes and complements, in which the former ones can be purchased instead of each other and the later ones can be purchased in addition to each other [13]. On one hand, understanding these relationships can help RS to generate more relevant candidates; on the other hand, they can also be applied in some specific scenarios, including "recommend after purchasing" and "similar products". Therefore there is a line of work [10,19] in distinguishing between substitute and complement. In particular, the complementary relationship between items is being studied in various product recommendation settings. For instance, in the grocery shopping scenario, Wan et al. [16] proposed Triple2Vec, which considers the cohesion of each triple pair (item, item, user) as item complementarity. Some recent studies also take time into consideration [17]. However, these existing methods mainly rely on co-occurrence signal or labels to acquire the semantics of inter-item relationship, which lacks theoretical support and assume static item-relationship. Instead, inspired by economics theory, our proposed network leverages price signal to quantify item complementarity and focuses on modelling the dynamics of item relationship, which better fit with the reality.

2.2 Graph-Based Neural Recommendation Models

With the emergence of neural graph embedding algorithms, graph structure has become a popular technique for various recommendation scenarios [8,22]. Different from the traditional latent factor model, high-order connectivity in the static graph can help generate enriched latent representations for users or items, which better capture both inter-user [3,18,20] and inter-item [22] relationships. Yet, existing methods do not integrate dynamics of relationships in recommender systems. To model dynamic patterns, on top of users and items, session information has been considered in the graph construction phase. For example, Session-based Temporal Graph (STG) [21] can effectively capture users' long-term and short-term preference via the random walk approach. Besides, there is also work [14] using RNN to capture dynamic user behaviour and a graph attention layer to model the social influence on a static-user graph. However, these methods are mainly designed for a specific community, which cannot be easily adapted to the task of next-period recommendation as stated in Subsect. 3.1.

3 Preliminaries

3.1 Problem Definition

Given a series of graph snapshots G_1, G_2, ..., G_t in different time intervals, G_t represents (V, E_t), which specifies the graph at time interval t. In the graph construction process, all user and item nodes in the data set are applied and edges E_t between nodes are also formulated. Since there are no overlapping time intervals, E_t and E_{t+1} are independent of each other and edges can be constructed and deconstructed between different pairs of existing nodes. In formal representation, given j graph snapshots in chronological order and target user u_i, we aim to recommend a set of items u_i feel interested in the time interval $j + 1$.

3.2 Cross Elasticity of Demand

According to the economics definition, Cross Elasticity of Demand describes the responsiveness in the quantity demand of one good when the price of another good changes [9]. Assume other factors remain unchanged, it depicts how the percentage change of price of one item might lead to the percentage change of quantity demand for another item. To ensure the computed percentage change is not adversely impacted by the change of absolute value, we take the reference of the arc elasticity of demand [9], and the equation is shown below:

$$E_{AB} = \frac{\frac{\Delta Q_{dA}}{\frac{Q_{dA1}+Q_{dA2}}{2}}}{\frac{\Delta P_B}{\frac{P_{B1}+P_{B2}}{2}}} = \frac{\Delta Q_{dA}}{\Delta P_B} \times \frac{P_{B1}+P_{B2}}{Q_{dA1}+Q_{dA2}} \tag{1}$$

where Q_{dA1} and Q_{dA2} are the quantity demand of item A at the initial time point and end time point, P_{B1} and P_{B2} are the price of the item B at the initial time point and end time point. In terms of its application, the computation result of cross elasticity of demand can be used to classify the relationship between any two goods. Substitutes have positive cross elasticity while complements have negative cross elasticity. The larger the absolute value of cross elasticity, the stronger the magnitude of its relationship kind. In our framework, following the practice of CDM [12], we only consider complementary item pairs with negative cross elasticity. To minimize the noise brought by weak item-item relationships, we locate the top-N strongest complementary item set of every single item by comparing the magnitude of computed cross elasticity.

4 Dynamic Complementary Graph Neural Network (DCGNN)

4.1 Framework Overview

DCGNN is a two-component recommendation framework based on a dynamic graph neural network, which integrates both dynamic user-item and item-item relations. The first component (Fig. 1), Complementary Graph Neural Network (CGNN), is a graph neural network to encode the structural information of user-purchase and temporal item complementarity (calculated in Sect. 3.2) into embeddings for static time frames. The second component is a time transfer mechanism (Fig. 2), which helps each time frame to inherit historical knowledge embedding from the previous time frame.

4.2 CGNN

Based on the user-item and item-item interaction relationship, we aim to get more effective collaborative signals for recommender systems via extraction of multi-hop connections in the graph, so CGNN is designed here. Firstly, similar to mainstream recommendation algorithms, CGNN projected every user/item node

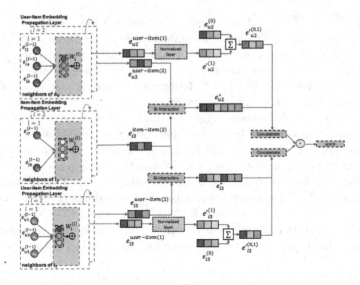

Fig. 1. Overview of Complementary Graph Neural Network: the goal is to acquire temporarily more expressive node embeddings for a static time frame by considering both user-item and item-item multi-hop information via graph neural networks.

to a high dimensional space. In Fig. 1, $e_{u2}^{(0)}$ and $e_{i3}^{(0)}$ represent the embedding of user and item respectively. After embedding initialization, instead of directly computing the user-item preference score in the prediction layer, CGNN adds information aggregation layers in-between. Within every layer, CGNN uses the sub-graphs to aggregate the relationships of user-item and item-item, so that more explicit multi-hop collaborative signals are used in node representation.

To explore multi-hop signals, we can stack more information aggregation layers, where the node representation can be shown below:

$$e_h^{(l+1)} = \frac{1}{\sqrt{|N_u||N_i|}} W_I e_h^{(l)} + \sum_{(h,r,t)\in N_h} \frac{1}{\sqrt{|N_u||N_i|}} W_r e_t^{(l)} \tag{2}$$

where N_u (user) and N_i (item) are one-hop directed graph node neighbours, l represents the number of aggregation layers, W_I represents the weight matrix of self-loop, W_r represents the weight matrix after aggregation of different types of relationships (e.g. user-item, item-item). In e^{l+1}, it has already incorporated the information of l-hop neighbours, which provide multi-hop neighbourhood information for every node. To facilitate batch training, we use the following equations to facilitate batch training for 2-hop neighbour aggregation.

$$E^{(l)} = (L + I)E^{(l-1)}W_1^{(l)} \tag{3}$$

$$L = D^{-\frac{1}{2}}AD^{-\frac{1}{2}} \tag{4}$$

$$A = \begin{bmatrix} 0 & R \\ R^{\top} & Q \end{bmatrix} \tag{5}$$

where $E^{(l)} \in \mathbb{R}^{N+M} \times d_l$ represents the user and item embeddings after l layers of propagation, $E^{(0)}$ represents embeddings at the initial state, L represents the laplacian matrix of user-item graph network, A represents the adjacency matrix, D represents the diagonal matrix, 0 represents all-zero matrix, $R \in \{0,1\}^{N \times M}$ and $Q \in \{0,1\}^{M \times M}$ represent the matrix containing all user-item and item-item interactions respectively.

Since one-hop and two-hop neighbour information play different roles in CGNN, we assign different fusion operations. For one-hop neighbour information, as previous work [4] has shown that historical user-item interactions can boost the performance of the ranking task, CGNN uses the one-hop user-item relationship to fine-tune the representation of the embedding by doing summation of initial embedding $e^{(0)}$ and the user-item relationship embedding $(e^{user-item(1)})$ after normalization (Eq. 9). For two-hop neighbour information, CGNN performs the bi-interaction operation for two-hop user-item information and item-item information to facilitate the interaction between two relationship features (Eq. 10).

$$e'^{(0,1)} = e^{(0)} + normalize(e^{user-item(1)}) \qquad (6)$$
$$e^* = (e^{user-item(2)} + e^{item-item(2)}) \oplus (e^{user-item(2)} \odot e^{item-item(2)}) \qquad (7)$$

Then CGNN combines e^* and $e'^{(0,1)}$ embeddings by doing concatenation and derives user-item preference score by performing dot product between the user and item prediction embedding.

$$e^{predict} = e'^{(0,1)} \odot e^* \qquad (8)$$
$$score = e_u^{predict} \cdot e_i^{predict} \qquad (9)$$

4.3   Time Transfer Mechanism

The main objective of the time transfer mechanism is to explore an effective knowledge distillation approach to inherit external/historical prior knowledge for the next time frame. In Fig. 2, CGNN trains the model by only considering user-item and item-item edges at time frame $t-1$. At the last training epoch, the model can give three intermediate embedding outputs, including: 1) embedding e_t^1 that encodes user preferences, 2) embedding e_t^2 that encodes the 2-hop user-item neighbours information, and 3) embedding e_t^3 that encodes the 2-hop item-item neighbours information. Consequently, to help the nodes to consider the knowledge from all three types of historical learning, the mechanism enables the knowledge inheritance by transferring the combined embedding as the initial embedding e_t^1 of CGNN at the next time frame t. The inherited embedding initialization can be formulated as equations below:

$$e_t^1 = f(E_{t-1}) \qquad (10)$$
$$E_t = (e_t^1; e_t^2; e_t^3) \qquad (11)$$

where $f(E_{t-1})$ can be different transfer mechanisms to inherit three different embeddings, which include:

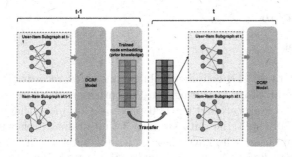

Fig. 2. Time transfer: how historical knowledge can be inherited to future periods.

- **SUM:** it means the summation of three historical knowledge embeddings: $f(E_{t-1}) = e_t^1 + e_t^2 + e_t^3$.
- **PRODUCT:** it means the product of three historical knowledge embeddings: $f(E_{t-1}) = e_t^1 * e_t^2 * e_t^3$.
- **MLP**: it means the prediction of future user preference: $f(E_{t-1}) = W_1 e_t^1$.
- **DIRECT:** it means the direct transfer of historical user preference embedding: $f(E_{t-1}) = e_t^1$.
- **ATTENTION:** it means the integration of three historical knowledge embeddings, where $\mathbf{Q}$ and $\mathbf{K}$ represent query and key projection matrix respectively: $f(E_{t-1}) = softmax([q^\top K e_1^t, q^\top K e_2^t, q^\top K e_3^t])E_{t-1}$, in which $q = (\mathbf{Q}e_1^t + \mathbf{Q}e_2^t + \mathbf{Q}e_t^3)/3$.

Furthermore, at the next time frame t, updated user-item and item-item edges will also be renewed in G_t. Through retraining the model parameters at time frame t, node representations involving relationship changes will be incrementally adjusted. Until n time frames, the trained node representation in G_n will incorporate the most accurate and update-to-date knowledge, and make predictions of a series of user's interested product in t_{n+1}.

4.4   Model Training

DCGNN adopts the pair-wise learning method. During the training process, we randomly sample 1 negative item for each positive item. Besides, we use the loss function below to maximize the difference between two samples.

$$L(Y, f(x)) = -log(\sigma(score_{pos} - score_{neg})) + L2 \tag{12}$$

where $e_u^{predict}$ represents prediction embedding of user, $e_i^{predict}$ represents prediction embedding of item, $score_{pos}$ and $score_{neg}$ represent the predicted user-item score for positive and negative samples respectively.

5 Experiment

5.1 Data Set

Order: This data set [17] contains mobile transaction records of multiple real-world offline retail stores[1]. During the 9-month period, there are more than 56,000 users who make around 920,000 purchases on more than 20,000 items. We chronologically split the data set into 3 time frames each with around 12 weeks. Also, only users who have made at least 1 purchase in all 3 time frames are kept. The first 2 time frames are the training set, the first two weeks of the 3rd time frame as the validation set and the following two weeks are the test set.

Dunnhumby[2]: Dunnhumby records transactions of 102 weeks from 2,500 households who do frequent shopping at a retail store. Households in total make more than 1.3 million purchases of over 37,000 items. For data pre-processing, we chronologically sort and split the data set into 6 time frames each with 17 weeks. Besides, we filter users and items with less than 5 interactions. Furthermore, we only retain items that have been purchased at least once in all 6 time frames. For data split, while the first 5 time frames are the training set, the 1st-2nd and 3rd-4th weeks of the 6th time frame are the validation and test set respectively.

5.2 Experimental Settings

Baselines: We compared DCGNN with following baselines: **BPR** [11]: A widely used pairwise learning method for item recommendations; **NCF** [5]: A deep learning method which uses users' historical feedback for item ranking. It combines **GMF** with a multilayer perceptron; (**MLP**), the two parts of NCF are also used as baselines; **GRU4Rec** [6]: A sequential recommendation model which utilizes GRU gate to compute ranking scores; **LightGCN** [4]: A state-of-the-art GCN-based recommendation model, which contains the most fundamental component of GCN-neighbourhood aggregation for collaborative filtering.

Evaluation Metrics: We adopt two popular metrics, Recall@K (Recall@K) and Normalized Discounted Cumulative Gain (NDCG@K), to evaluate the performance of top-N recommendation.

Implementation Details: We implement the model in Pytorch. To ensure fairness, the embedding size is set to 64 for all models. All the hyper-parameters are tuned to get the best result in the validation data set. For GRU4Rec, due to different time duration, we consider purchase history length of 10 and 20 for the Order and Dunnhumby data sets respectively. For LightGCN, we incorporate both user-item purchase relations and item complementarity in the graph construction process. For DCGNN, main parameters include top-N (10) complementary items, learning rate (1e−3) and L2 regularization (1e−5). All parameters are initialized with 0 mean and 0.01 standard deviation.

---

[1] https://github.com/THUwangcy/SLRC/tree/master/data.
[2] https://www.dunnhumby.com/sourcefiles.

6   Experimental Results and Analysis

6.1   Performance Comparison

From Table 1, our proposed model DCGNN performs significantly better than all baselines in both data sets, which showcases its effectiveness in two ways. First, it incorporates multi-hop neighbours information of both user-item purchase and item-item complementarity relationship through graph neural networks. Second, it effectively incorporates incrementally updated historical knowledge into the model via the time transfer mechanism.

Table 1. Test results on two data sets. We repeat the experiment five times with different random seeds. The average results are reported. Best Baselines are underlined, ** means significantly better than the strongest baselines ($p < 0.01$)

Method	Order				Dunnhumby ,			
	Recall@5	Recall@10	NDCG@5	NDCG@10	Recall@5	Recall@10	NDCG@5	NDCG@10
MLP	0.0592	0.0899	0.0393	0.0492	0.1518	0.2189	0.1055	0.1271
GMF	0.0415	0.0662	0.0271	0.0351	0.1685	0.2476	0.1134	0.1389
NCF	0.0516	0.0892	0.0334	0.0455	0.1555	0.2209	0.1082	0.1293
BPR	0.0286	0.0468	0.0185	0.0243	<u>0.1935</u>	<u>0.2663</u>	<u>0.1376</u>	<u>0.1612</u>
LightGCN	0.0585	0.0904	0.0391	0.0494	0.1428	0.2035	0.0987	0.1182
GRU4Rec	<u>0.0706</u>	<u>0.1097</u>	<u>0.0464</u>	<u>0.0590</u>	0.1285	0.1781	0.0891	0.1051
DCGNN	**0.0953****	**0.1484****	**0.0614****	**0.0785****	**0.2260****	**0.3023****	**0.1613****	**0.1859****

For baseline methods, in the Order data set, GRU4Rec performs the best among all baselines due to its good fit with the data set scenario. In grocery shopping settings, users tend to shop more regularly with a specific pattern to fulfil daily needs. Therefore the change of user intent becomes a critical factor for model performance. With a shorter purchase history per user, the history length of 10–20 is sufficient for GRU4Rec to fully capture the sequential transition of the purchased items. Second, comparing LightGCN with the rest of the baselines, it performs better due to its ability to incorporate multi-hop neighbour information via multiple layers.

In the Dunnhumby data set, first, BPR performs the best among all baselines due to the high data density of the data set. Different from the Order data set, the Dunnhumby data set only contains a relatively small number of loyal and frequent-buying members. Therefore, CF-based methods like BPR are advantageous under highly dense data circumstance. Second, there is a difference between the performance of DCGNN and LightGCN despite sharing the same nature of the CF-based method. The main reason is that LightGCN tends to treat inter-node relationships equally in all timeframes, which makes it fail to capture the dynamics of item-item complementarity along different periods. In contrast, DCGNN puts a stronger emphasis on considering inter-node relationships in the latest timeframe during the prediction phase, which ensures the node representation is always in its best shape before prediction. Third, to explain the low performance of GRU4Rec, it is mainly due to the long purchase history per user in the Dunnhumby data set. As GRU4Rec only relies on the sequential transition of purchased items to learn user representation, the history length considered by the algorithm can only reflect a small portion of users' purchase history. Therefore, the inability to accurately express users heavily undermines the recommendation performance of GRU4Rec.

Table 2. Comparison between "with transfer" and "without transfer" mechanism

Transfer mechanism	Order		Dunnhumby	
	Recall@5	NDCG@5	Recall@5	NDCG@5
Without transfer	0.0834	0.0527	0.2118	0.1506
With transfer	**0.0953**	**0.0614**	**0.2260**	**0.1613**

6.2 Effect of Capturing Dynamic Item Complementarity

To examine the effect of the time transfer mechanism, we perform an ablation study on the time transfer mechanism. Specifically, we try to train the model with CGNN only. In Order and Dunnhumby data sets, we initialize all node embeddings with 0 mean and 0.01 standard deviation, and only consider the user-item purchase and item complementarity relations at the last training time-frame (the 2nd timeframe in the Order data set and the 5th timeframe in the Dunnhumby data set). Then, the model makes predictions on the same test set. From Table 2, we observe that DCGNN with time transfer mechanism gets a better recommendation performance under both Recall@5 and NDCG@5 evaluation metrics. It is because while historical knowledge of item complementarity is useful for users' preference prediction, the time transfer mechanism can also enable DCGNN to effectively distillate external prior knowledge in the form of embedding in consecutive transfers during the training process.

6.3 Effect of Transfer Mechanism

In total, there are three knowledge embeddings, which encode information of user preference, users' neighbour preference and item-item complementarity respectively. For time transfer mechanism, we explore various approaches to help the proposed model inherit historical knowledge embeddings from the previous time frame. From Table 3, we notice the performance vary significantly across different data sets, and the optimal choice is not consistent. DIRECT generally gets promising result, but ATTENTION is more powerful in the Order data set. To explain the variance, for Order data set, ATTENTION performs better because it effectively combines three knowledge historical embeddings by different weightings. For Dunnhumby data set, as it is a much dense data set with far fewer users, user interests are more consistent along different time frames, thereby incorporating multi-hop user-item and item-item knowledge might disrupt its user preference representation and lowers its recommendation performance.

Table 3. Comparison between different transfer mechanisms

Transfer mechanism	Order		Dunnhumby	
	Recall@5	NDCG@5	Recall@5	NDCG@5
PRODUCT	0.0802	0.0503	0.2075	0.1477
SUM	0.0866	0.0558	0.2255	0.1608
MLP	0.0885	0.0563	0.2195	0.1570
DIRECT	0.0936	0.0598	**0.2260**	**0.1613**
ATTENTION	**0.0953**	**0.0614**	0.2188	0.1566

7   Conclusion and Future Work

Modelling Dynamic item complementarity is practical for real-world recommendation scenario. To consider the time-sensitive nature of relationships, we proposed a Dynamic Complementary Graph Neural Network (DCGNN) to integrate the dynamic item-complementarity into recommendation systems.

In both real-world data sets, experiment results show DCGNN can effectively model dynamic item complementarity and boost the performance of the recommender system. In other applications, DCGNN's recommendation list can also be applied in a wide variety of personalized sales channels, such as email, message or updates from mini-program, to attract more customers to make purchases. In the future, we plan to introduce more theory in economics, psychology, or other disciplines to enhance current recommendation algorithms, which can improve the algorithm performances and support the designing of new methods.

Acknowledgements. This work is supported by the National Key Research and Development Program of China (2018YFC0831900), Natural Science Foundation of China (Grant No. 62002191, 61672311, 61532011) and Tsinghua University Guoqiang Research Institute.

References

1. Covington, P., Adams, J., Sargin, E.: Deep neural networks for YouTube recommendations. In: Recsys 2016, pp. 191–198 (2016)
2. Dong, Y., Chawla, N.V., Swami, A.: metapath2vec: scalable representation learning for heterogeneous networks. In: SIGKDD 2017, pp. 135–144 (2017)
3. Fan, W., et al.: Graph neural networks for social recommendation. In: The Web Conference, pp. 417–426 (2019)
4. He, X., Deng, K., Wang, X., Li, Y., Zhang, Y., Wang, M.: LightGCN: simplifying and powering graph convolution network for recommendation. arXiv preprint arXiv:2002.02126 (2020)
5. He, X., Liao, L., Zhang, H., Nie, L., Hu, X., Chua, T.S.: Neural collaborative filtering. In: The Web Conference, pp. 173–182 (2017)
6. Hidasi, B., Karatzoglou, A., Baltrunas, L., Tikk, D.: Session-based recommendations with recurrent neural networks. arXiv preprint arXiv:1511.06939 (2015)

7. Kang, W.C., Wan, M., McAuley, J.: Recommendation through mixtures of hetero-geneous item relationships. In: CIKM, pp. 1143–1152 (2018)
8. Ma, W., et al.: Jointly learning explainable rules for recommendation with knowl-edge graph. In: The World Wide Web Conference, pp. 1210–1221 (2019)
9. Marshall, A.: Elements of Economics of Industry: Being the First Volume of Ele-ments of Economics, vol. 1. Macmillan (1898)
10. McAuley, J., Pandey, R., Leskovec, J.: Inferring networks of substitutable and complementary products. In: SIGKDD, pp. 785–794 (2015)
11. Rendle, S., Freudenthaler, C., Gantner, Z., Schmidt-Thieme, L.: BPR: Bayesian personalized ranking from implicit feedback. arXiv preprint arXiv:1205.2618 (2012)
12. Shao, Y., Zhang, M., Ma, W., Wang, C., Liu, Y., Ma, S.: Integrating latent item-item complementarity with personalized recommendation systems. J. Softw. **31**(4), 1090–1100 (2020)
13. Shocker, A.D., Bayus, B.L., Kim, N.: Product complements and substitutes in the real world: the relevance of "other products". J. Mark. **68**(1), 28–40 (2004)
14. Sun, P., Wu, L., Wang, M.: Attentive recurrent social recommendation. In: SIGIR, pp. 185–194 (2018)
15. Vasile, F., Smirnova, E., Conneau, A.: Meta-prod2vec: product embeddings using side-information for recommendation. In: RecSys, pp. 225–232 (2016)
16. Wan, M., Wang, D., Liu, J., Bennett, P., McAuley, J.: Representing and recom-mending shopping baskets with complementarity, compatibility and loyalty. In: CIKM, pp. 1133–1142 (2018)
17. Wang, C., Zhang, M., Ma, W., Liu, Y., Ma, S.: Modeling item-specific temporal dynamics of repeat consumption for recommender systems. In: The World Wide Web Conference, pp. 1977–1987 (2019)
18. Wang, J., Ding, K., Zhu, Z., Zhang, Y., Caverlee, J.: Key opinion leaders in rec-ommendation systems: opinion elicitation and diffusion. In: WSDM, pp. 636–644 (2020)
19. Wang, Z., Jiang, Z., Ren, Z., Tang, J., Yin, D.: A path-constrained framework for discriminating substitutable and complementary products in e-commerce. In: WSDM, pp. 619–627 (2018)
20. Wu, L., Sun, P., Fu, Y., Hong, R., Wang, X., Wang, M.: A neural influence diffusion model for social recommendation. In: SIGIR, pp. 235–244 (2019)
21. Xiang, L., et al.: Temporal recommendation on graphs via long-and short-term preference fusion. In: SIGKDD, pp. 723–732 (2010)
22. Ying, R., He, R., Chen, K., Eksombatchai, P., Hamilton, W.L., Leskovec, J.: Graph convolutional neural networks for web-scale recommender systems. In: SIGKDD, pp. 974–983 (2018)

NLP for IR

LDA-Transformer Model in Chinese Poetry Authorship Attribution

Zhou Ai, Zhang Yijia, Wei Hao, and Lu Mingyu[(⊠)]

College of Information Science and Technology, Dalian Maritime
University, Dalian 116026, China
lumingyu@dlmu.edu.cn

Abstract. Authorship attribution broadly is defined as an analysis of individuals' writing styles, which has been attracting a lot of interest. Although the problem has been widely explored, no previous studies attempt to identify Chinese classical poetry. In this paper, we presented a public classical poetry corpus in Tang Dynasty for Chinese authorship attribution. As a particular literal form, the theme feature plays a crucial role in Chinese poetry authorship attribution. To integrate the topic feature of the Chinese poem, we employed the latent Dirichlet allocation model to capture the extra theme information. Meanwhile, due to the incoherent expression of poetry text, it is hard to capture incoherence information effectively from Chinese poems. To tackle this problem, we propose a combination model called LDA-Transformer to perform authorship attribution of Chinese poetry. We conduct systematical evaluations for the proposed method on three Chinese poetry datasets. The experimental results suggest that the topic feature can effectively improve the performance of authorship attribution in Chinese poetry. Our model achieves state-of-the-art results on related baseline methods.

Keywords: Authorship attribution · Chinese classical poem · Transformer · LDA

1 Introduction

Authorship attribution is a unique task that is closely related to both the representation of individuals writing style and text categorization [1]. The rationale behind this problem suggests that the linguistic structure of the documents can be reliably inferred from individual writing activities, which reflects their stylistic "fingerprint" unconsciously. Authorship attribution (AA) aims to determine the authors of a document among a list of candidates, which plays an essential role in many applications, including forensic investigation [2], terrorist identification [3] and the field of network security [4]. The task has been extensively studied among a wide range of languages. However, the research of Chinese AA is still in the early stages. So far, there is no public standard corpus for Chinese AA study. The most popular Chinese corpus for AA is The Dream of Red Mansion [5].

As a special literary form, classical poetry, especially for Chinese poetry in Tang Dynasty, not only had high artistic merit and appreciation value during that period, but

© Springer Nature Switzerland AG 2021
H. Lin et al. (Eds.): CCIR 2021, LNCS 13026, pp. 59–73, 2021.
https://doi.org/10.1007/978-3-030-88189-4_5

influenced Chinese culture and history afterward. Expect for some 'Yuefu Poems', like 'Song of a Pipa Player' and the 'Everlasting Regret', most poems are short (less than 50 characters). In majority, classical poems express rich implications with the streamlined script, besides polysemous is a pervasive phenomenon in this literary form. Simultaneously, Chinese classical poems have more restrictions on the number of characters, lines as well as the tonal styles. The most obvious features of classical poetry are different themes. Generally speaking, Gao Shi and Cen Shen are the representatives of frontier poets, while Wang Wei and Meng Haoran prefer to write pastoral landscape poems. Therefore, the themes are valuable for AA in poetry. Different from other short texts like tweeters, the expression of poetry is incoherent in time and space or even in grammar structure. For example, "Cock crow(s), thatched inn, moon; human trace(s), wood(en)bridge, frost." ("鸡声 茅店 月，人迹 板桥 霜"). The omission of verbs and prepositions causes the incoherent arrangement of nouns, which reflects the artistry of the poem as a whole. Hence, it is hard to capture incoherent information features and grasp the writing style of poetry in the mass.

Usually, the existing solutions in previous AA studies typically consist of three major steps, including the input documents, the traditional features process, and the machine learning classifications. In this paper, we captured the extra topic features through the latent Dirichlet allocation model (LDA) to identify the poets more effectively. Due to the incoherence of poetry texts, we employed the Transformer model to capture deep and incoherence information of the poems.

More specially, our major contributions are summarized as follows. Firstly, a public corpus of classical poetry for Chinese AA is established, called QuanTangShi Corpus. Secondly, we proposed a novel LDA-Transformer model to capture incoherence information for poetry AA, which is the first work attempting to integrate deep learning models for Chinese AA. Thirdly, the poem themes are integrated to improve AA, and achieve the state-of-the-art performance on three datasets.

2 Related Works

The studies on AA can be traced back over a hundred years. The first attempt of this area was based on statistical methods, which made a statistical analysis of word length distribution to identify Shakespeare's works [6]. Machine learning approaches have been successfully applied in AA [7, 8]. For example, random forests are able to effectively handle high-dimensional data, which has been widely used for AA. Some studies have shown that the results of Naive Bayes are also promising in AA [9]. Currently, the deep learning models have been proposed for short text AA, and achieved an AUC of 0.628 [10]. Similarly, in 2017, Shrestha [11] applied CNN models on the datasets of tweeters, the highest accuracy of 50 authors is 0.761.

For Chinese AA, the single most dominant issue is whether the last 40 chapters of the Dream of the Red Mansion are written by the same author as the first 80 chapters [8] from 1987 [12] till now. Some researchers tended to focus on other Chinese modern literary masterpieces like Martial arts novels of Louis Cha and Gulong [13] and prose [14]. So far, thereis few studies in AA on classical Chinese poetry. Despite some traditional features, namely, common words [15], punctuation [16], and N-gram [17], people also concern with special features in language domains. For example, Chinese auxiliary words [18] and the rimes of Chinese syllables (PinYin) [19]. Recently, Chinese classical poetry has attracted more attention in some natural language processing (NLP) domains, such as style modeling [20] and poetry generation [21].

3   Corpus

To establish the first corpus of classical poetry for Chinese AA, namely, QuanTangShi Corpus, we collected Chinese classical poetry of Tang Dynasty. However, the primitive data has many problems, such as duplication, errors and some random codes. We pre-processed the poetry data, including removing repetition, correcting the errors and random codes, separating the records. There are 905 poems written by unknown poets in the poetry data. Since such records cannot be applied to AA task, we deleted all these poems.

We annotated the poems with author and title, AA is regarded as the supervised learning task, so we treat the authors as the labels of the documents. The text files are organized into JSON documents. The annotation model is: QuanTangShi Model = [Author, Title, Poem]. The following is an example for a poem annotation: [Author: '孟浩然' "Meng Haoran", Title: '春晓' "Spring Morning", Poem: '春眠不觉晓，处处闻啼鸟，夜来风雨声，花落知多少？' "This spring morning in bed I'm lying, not to awake till birds are crying. After one night of wind and showers, how many are the fallen flowers?"].

Table 1. Corpus statistic

Poem number per author n	Author number	Total poem number	Average poem number
0<n<10	1954	3754	2
10<n<50	197	4512	23
50<n<100	42	3033	72
100<n<300	67	11298	169
300<n <500	23	8368	364
500<n<1000	15	9478	632
n>1000	2	4294	2147

Finally, the clean corpus totally contains 44734 poems created by 2300 authors, almost 19 poems per author. Table 1 illustrates the basic statistic information of the final corpus. It can be seen that the corpus is imbalanced. Nearly half of the poets in Tang Dynasty only produce 1 or 2 poems. The number of authors who create over 20, 50, and 100 poems decrease dramatically to 251, 149 and 107 respectively. Among all the poets in Tang Dynasty, Bai Juyi is the most productive poet created 2844 poems in his life.

4 Methodology

In this section, we first give a brief introduction to our LDA-Transformer combination model. Then we describe this hybrid model in detail.

4.1 Model Architecture

As a special literal form, poems have not only plenty of incoherence information, but also some integral information which is valuable. For example, "With wine of grapes the cups of jade would glow at night" ('葡萄美酒夜光杯'), the character "grapes" ('葡萄') cannot be separated. What's more, topic information is an effective special feature for AA in Chinese classical poetry. To employ the advantage of topic information and the Transformer framework, we integrated them together and proposed the LDA-Transformer model for AA.

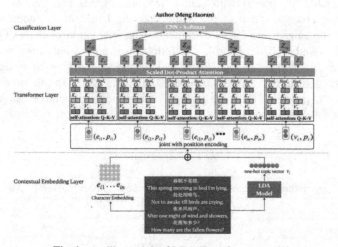

Fig. 1. An illustration of LDA-Transformer model

As demonstrated in Fig. 1, our framework contains a contextual embedding layer, a Transformer layer and a classification layer. For the contextual embedding layer, the LDA model is selected to extract the critical theme features, which is represented as a one-hot topic vector (Vi). Then, we implement the same multi-head attention layer, as the classic Transformer [22]. Finally, CNN instead of a fully connected network (FNN) was used for classification, which can acquire not only some indivisible features but also some long-range contextual information of a poem.

4.2 LDA Model for Topic Feature

As a special literary form, the theme of Chinese classical poetry is valuable for AA. In general, Chinese classical poetry can be divided into many subjects such as frontier, pastoral landscape, feminine querimony, farewell, history, and so on. Considering that there is no corpus labeled poem subjects, LDA is chosen to cluster poems.

As LDA is an unsupervised technique, one problem is how to determine the number of topics. In Blei's opinion [23], the most commonly used evaluation for the LDA model is the perplexity. More formally, the perplexity is:

$$perplexity(D_{test}) = exp\left\{ -\frac{\sum_{d=1}^{M} \log p(w_d)}{\sum_{d}^{M} N_d} \right\} \tag{1}$$

where M represents a test set of documents and N_d delegates the size of the document d (i.e., the number of the words), and we generalize the $p(w_d)$ as:

$$p(w_d) = \sum_z p(z)p(w|z, gramma) \tag{2}$$

where z means topic and w indicates the document and *gramma* is the distribution of document-topic from the training set. Consequently, the perplexity, used by convention in language modeling, is monotonically decreasing in the likelihood of the test data, and is algebraic equivalent to the inverse of the geometric mean per-word likelihood. A lower perplexity score indicates better. Such a measure is useful for evaluating the predictive model, but does not address the more exploratory goals of topic modeling.

However, there is an interesting twist here. The mathematically rigorous calculation of model fit (data likelihood, perplexity) doesn't always agree with human opinion about the quality of the model, as shown in a well-titled paper "Reading Tea Leaves: How Humans Interpret Topic Models" [24]. Therefore, topic coherence has been used to measures how often the topic words appear together in the corpus and reflects the degree of semantic similarity between high scoring words in the topic. Topic coherence can help distinguish between topics that are semantically interpretative topics and topics that are artifacts of statistical inference.

UMass Measure. There are two coherence measures designed for LDA, and both of them have been shown to match well with human judgments of topic quality: the UCI measure [25] and the UMass measure [26]. Both of them compute the coherence score C as the sum of pairwise scores on the set of the words V used to describe the topic. We generalize this as:

$$C(V) = \sum_{(v_i, v_j) \in V} score(v_i, v_j, \varepsilon) \tag{3}$$

where V is a set of topic words and ε indicates a smoothing factor, which guarantees that score returns a real number. (Usually, we would like to select $\varepsilon = 1$).

There is no appropriate external corpus for Chinese classical poetry computing the word probabilities. We choose the UMass who measure the score based on document co-occurrence:

$$score(v_i, v_j, \varepsilon) = log\frac{D(v_i, v_j) + \varepsilon}{D(v_i)} \tag{4}$$

where $D(x, y)$ counts the number of documents containing words x and y. $D(x)$ counts the number of documents containing x.

4.3 Transformer

Classical Transformer has an encoder-decoder structure. As AA is a classification task, we only use an encoder structure. As shown in Fig. 1, in order for the model to make use of the order of the sequence, we add positional encodings to the input embeddings at the bottoms of the Transformer model. Then instead of performing a single attention function, we found it is beneficial to linearly project the queries, keys and values h times with different, learned linear projections. As demonstrated in Fig. 2, mapping Q, K, V though h different lineal transformations getting an array of attentions, different attentions focus on different information. The output of each head attention is concatenated and once again projected, producing the final values.

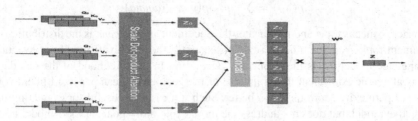

Fig.2. Multi-head attention

The crucial part of multi-head attention is the scaled dot-product attention which can be implemented using a highly optimized matrix multiplication operation. Comparing with the most common one, additive attention [27], and multiplicative attention is much faster and more space-efficient. In Vaswani's opinion, the input consists of queries and keys of the dimension d_k, and values of the dimension d_v. We compute the dot products of the query packed together into a matrix Q with all keys packed together into a matrix K, divide each by $\sqrt{d_k}$, which can make the gradient update more stable and apply a softmax function to obtain the weights on the values packed together into a matrix V. We calculate the matrix of outputs as:

$$Attention(Q, K, V) = softmax\left(\frac{QK^T}{\sqrt{d_k}}\right)V \qquad (5)$$

Finally, for each poem we adopt a multi-scale CNN model to predict the author label. With the help of multiple filters (usually chose 1,2,3,4 size), CNN can acquire context information from various dimensions accurately and effectively, especially for some in coherence information of a poem. Then we apply a max-pooling layer to capture the most essential features. Finally, these features are passed to a fully connected softmax layer whose output is the probability distribution over labels.

5 Experiments

In this section, we compared the performance of our LDA-Transformer with several baselines. A set of common metrics were adopted to evaluate our model: Accuracy, Precision, Recall, and F1- score. Meanwhile, we describe the datasets and show some visualization comparison results and analyse the experimental results in detail.

5.1 Datasets

We evaluated our model on three datasets. The first dataset (LD) includes LiBai and DuFu's poems. Both of LiBai and DuFu are the most shining stars not only in Tang Dynasty, but also in the whole development of Chinese culture. The second dataset (WYLL) includes the poems written by the four-talented poets in the early Tang Dynasty. The third dataset (12 Poet) collects of 12 poets delegated different periods of Tang Dynasty. Most of them are very famous and have produced more than 300 poems, Table 2 illustrating the basic information of the 12 Poet dataset. These datasets have a different number of authors and document sizes, which allows us to perform experiments and tests our approach in different scenarios.

For all datasets, 80% of them are used for training, others for testing. Since none of the datasets have a standard development set, we randomly select 10% of the training data for this purpose. Early stopping is used on the development sets and Adam with shuffled mini-batches (batch size 16) is used for optimization. To avoid overfitting, 25% dropout and L2 regularization are used. The optimization objective is standard cross-entropy errors of the predicted character distribution and the actual one. Table 3 shows descriptive statistics for the datasets.

5.2 Baseline

We consider the following state-of-the-art AA deep learning models and some popular machine learning models for comparison:

Naive Bayes: Yi [28] first applied this model to AA in Chinese classical poetry and achieved exciting results in binary classification.
SVM: Stamatatos E. [7] suggests that SVM is the most effective model, especially for long literal texts AA.
CNN: CNN is an effective deep learning model for AA [11], which achieves high performance in short texts.
BERT: BERT [29] obtains new SOTA results on eleven NLP tasks. In this paper, we use the pre-trained BERT-Based Chinese for evaluation.

5.3 Experimental Results

Effect of the Parameter. As LDA is an unsupervised model aiming to find the optimal number of topics, we built different LDA models with different values of the number of topics (k) and picked the one that gives the highest coherence value. Choosing a 'k' that marks the end of a rapid growth of topic coherence usually offers meaningful and explicable topics. Figure 3 below shows the changing trend of coherence scores with the increasing number of topics (k) on different datasets (i.e., LD, WYLL, 12 Poet).

Results Table 4 presents the performance on the three selected datasets. From the experimental results, we have the following observations.

Intuitively, the effect of the deep learning model is better than that of machine learning. Meanwhile, compared with the previous cognition, SVM is the most effective

Table 2. 12 poet dataset.

Period	Author	Poems
Early	Wang Bo	89
	Luo Binwang	130
	Lu Zhaolin	106
Prosperous	Li Bai	957
	Du Fu	1450
	Wang Wei	382
Mid	Bai Juyi	2844
	Yuan Zhen	800
	'Liu Yuxi'	797
Late	Li Shangyin	577
	Du Mu	527
	Wen Tingyun	353

Table 3. Datasets statistics

Dataset	LD	WYLL	12 Poet
Authors	2	4	12
Poems	2407	358	9012
Train	1684	250	6308
Dev	241	36	902
Test	482	72	1802
Average poems	1204	90	751
Total characters	211484	28581	708616
Average characters	88	80	79

classifier; Naive Bayes acquires a higher accuracy in poetry text, even higher than the basis of CNN, in terms of binary classification. According to recent studies, BERT has achieved SOTA performance in some NLP tasks. But our experiments take the opposite result. For most of our datasets the performance barely satisfactory, only a few of them get almost the same accuracy as CNN, others far behind the basis of CNN, let alone our model. The most probable reason is the ancient Chinese is very different from modern

Chinese. The BERT-Base Chinese model pre-trained by modern Chinese, which is not applicable to ancient Chinese like classical poetry.

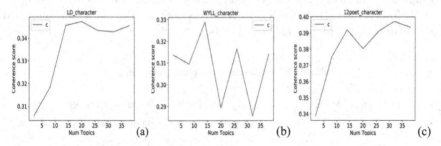

Fig.3. Effectiveness of topic coherence on three datasets, (a) for LD, (b) for WYLL, (c) for 12 poet

Table 4. Experimental results on three poetry datasets

Corpus	Model	Accuracy	Precision	Recall	F1-score
LD	NB	91.49%	91.62%	91.49%	91.53%
	SVM	88.38%	88.60%	88.38%	88.44%
	CNN	91.49%	91.50%	91.49%	91.50%
	BERT	91.49%	91.49%	91.90%	91.50%
	Ours	**94.40%**	**94.42%**	**94.40%**	**94.41%**
WYLL	NB	66.67%	66.67%	66.67%	66.67%
	SVM	61.11%	61.11%	61.11%	61.11%
	CNN	70.83%	70.83%	70.83%	70.83%
	BERT	72.22%	73.12%	72.22%	72.22%
	Ours	**79.17%**	**81.37%**	**79.17%**	**76.92%**
12 Poet	NB	61.71%	62.00%	61.71%	61.71%
	SVM	57.99%	58.34%	57.99%	57.95%
	CNN	62.16%	61.16%	62.16%	60.95%
	BERT	62.00%	62.15%	62.75%	61.16%
	Ours	**66.26%**	**64.87%**	**65.14%**	**65.54%**

The proposed model LDA-Transformer achieves the best performance on three datasets, especially in terms of Accuracy and F1-scores, our model gains from 5.6% to 1.5% improvement among all datasets. The results strongly demonstrate the effectiveness of our proposed LDA-Transformer framework.

5.4 Ablation Study

To illustrate the validity of three components, the corresponding evaluation is made in this subsection. In this experiment, we test the three simplified models by dropping the LDA, the Transformer and the CNN component, respectively. Then we test the simplified model on the LD, WYLL and 12 Poet datasets.

Table 5. Effectiveness of different components

Corpus	Method	Accuracy	Precious	Recall	F1-score
LD	Ours	**94.40%**	**94.42%**	**94.40%**	**94.41%**
	-No LDA	93.57%	93.58%	93.57%	93.57%
	-No CNN	91.91%	91.49%	91.98%	91.90%
	-No Transformer	91.49%	91.50%	91.49%	91.50%
WYLL	Ours	**79.17%**	**81.37%**	**79.17%**	**76.92%**
	-No LDA	73.61%	76.48%	73.61%	73.70%
	-No CNN	73.61%	74.42%	74.21%	73.61%
	-No Transformer	70.83%	70.83%	70.83%	70.83%
12 Poet	Ours	**66.26%**	**64.87%**	**65.14%**	**65.54%**
	-No LDA	65.54%	64.73%	64.14%	64.16%
	-No CNN	63.93%	63.60%	63.93%	62.75%
	-No Transformer	62.16%	61.16%	62.16%	60.95%

Table 5 suggests that LDA structure plays a relatively important role in our model. Especially for the WYLL dataset, it acquires the best performance which is increased by 5.56% in terms of accuracy. We can also find that the Transformer framework contributes the most to our model (2.91% increment for LD, 8.34% for WYLL and 4.10% for 12 Poet of the accuracy). The improvement from our model on the three selected datasets in terms of F1-scores is statistically significant. Besides, the performance of -No Transformer model is worst on 12 Poet datasets in terms of F1-scores. We suggest that the dataset of 12 Poet needs to improve the effectiveness of capturing incoherence information than the other two datasets because of the increasing number of the authors and the epoch differences. Hence only adopt CNN model cannot effectively extra incoherence information in a poem. The results illustrate that the Transformer framework is a significant component in our model.

Similarly, it can be observed that multi-scaled CNN model enhances the performance of our model, which contributes 2.49%, 5.56%, 2.33% of accuracy on the LD, WYLL and 12 Poet datasets, respectively. Thus, CNN classification is an important component of our model. Besides we also see that as the number of authors increases, the performance of the Transformer model is fallen dramatically. Therefore, both the Transformer framework and the CNN model are vital parts of the proposed model.

5.5 Visualization

In this section, we indicate our LDA-Transformer model not only capture the long-range incoherence information more effectively, but also wholly grasp the writing style of Chinese classical poetry in Tang Dynasty by visualization.

The multi-head attention layer was visualized in Fig. 4 for a 'Jueju' created by Du Fu as an example. We separately generate the long-range left and right character embeddings in a poem by multi-head attention with a residual connection. Figure 4 (a) represents one-layer multi-head attention visualization and Fig. 4 (b) draws a picture of the 6 layers multi-head attention with some details inside.

Both of the two plots indicate the effect of our LDA-Transformer model. Many of the head's attention to the long-range dependence of the character '花' "flowers". Figure 4 (a) shows that when there is only one layer, the proposed model can capture the incoherence information from the first two sentences of the poem. As the layer becomes deeper, the model starts to capture contextual information nearby, as illustrated in Fig. 4 (b). Hence our model can effectively capture incoherence information and make the correct decision.

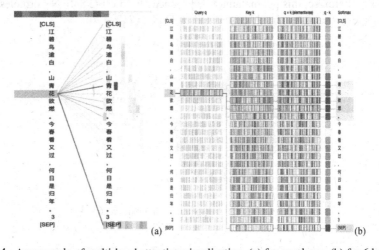

Fig. 4. An example of multi-head attention visualization, (a) for one-layer, (b) for 6 layers.

6 Error Study

In previous sections, a set of experiments has been shown through visualization of how our LDA-Transformer model can capture incoherence information and use them to identify the actual author of the poems. There are cases where the model fails its task and generates the wrong output. Knowing what causes the model to fail is of much interest, as it reveals the limitations of our models and helps improving future designs. In this section, we perform an error analysis on the 12 Poet dataset. We categorized the failure cases into four major groups. The overall share of each error category is shown in Fig. 5.

Although the results of the error analysis conducted in this session are only for 12 Poet dataset, similar causes are the source of errors for other datasets.

It is worth mentioning that not all cases are 100% distinct from each other and there may be the possibility of overlaps for some failure cases meaning that one sentence is misclassified as the result of multiple causes.

6.1 Contradiction

The existing proper nouns in the short poems or in the poems' titles can drop a hint for AA. However, our proposed model LDA-Transformer ignores presentation leading to making contradictory decisions.

As a human, no one can be alive through all periods of Tang Dynasty (618–907), so we separate the poets into four periods. However, the examples of failures show that the actual author often identified as a poet living in another period of Tang Dynasty. Real age of a poem can be attributed to the proper nouns, especially the names in the poem or in the title. For example, a poem created by Wen Tingyun whose title is 'Reply Prime Minister Linghu', here, Prime Minister Linghu represent Linghu Tao, who become prime minister at BC 850. Therefore, this poem cannot be produced before BC 850. Nevertheless, our model suggests the poem should be created by Du Fu (712–770).

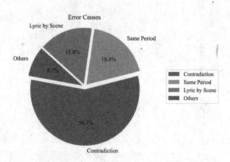

Fig. 5. Distribution of different error causes

In Tang Dynasty it is common that many poets like to reply others' poems, namely, 'Heshi' or 'Zengshi', which can be easily found in the titles. For example, 'Reply to Bai Juyi as a Gift for First Time at Banquet in Yangzhou' written by Liu Yuxi. Common sense suggests that the author of this poem is impossible Bai Juyi. The title has already given the information of the true author. However, our model fails in this case.

The contradiction is the most common cause of failures as shown in Fig. 4. This is responsible for over half of failures showing that further improvements of the model performance highly require addressing this issue.

6.2 Same Period

As poems belong to ancient Chinese with slow language evaluation, most poets living in the same period of Tang Dynasty share similar characters, common words or even tonal styles. Hence, even if more features like words, the rimes of Chinese syllables (PinYin) and tonal styles are adopted, this kind of error proves to be the new challenge for poetry AA in the future.

Although some error poems also exist proper nouns, due to these poets living in the same period, even some of them are friends. For example, Yuanzhen, LiuYuxiand Bai Juyi who live in the same period and share a similar experience of life. Therefore, if the wrong poets stay in the same period with the correct ones, it is challenging for a human to distinguish by existing proper nouns, let alone by neural network models.

From Fig. 5 the same period mistake is responsible for 18.8% of failures. Only employs more traditional features cannot make a distinction between correct poets and the error ones. Therefore, this type of error will be difficult for poetry AA from now on.

6.3 Lyric by Scene

The third group of errors happens to the lyric by scene poems. Generally speaking, these types of poems rarely have an obvious sign indicating years and usually describe similar contents for scenery, like 'Autumn Evening in the Mountains' created by Wang Wei. Only use character features can hardly make correct decisions. We need to fuse other features for future improvement.

There are also a few numbers of errors where the causes do not fall into the existing categories. In some cases, there is more than one reason causes the failures, or it might be the case where the visualization is not able to capture the cause of failure. These cases are shown in Fig. 8 like others.

7 Conclusions

In this paper, an LDA-Transformer model was proposed for AA and it shows considerable performance in poetry text. To the best of our knowledge, we are the first effort to use Chinese poems of Tang Dynasty as a corpus for AA. Besides, our proposed model can effectively capture incoherence information and grasp the writing style of classical poems. In addition, as a special literal form, the theme of the poetry does improve the accuracy of poets' attribution. The experimental results show that our proposed model achieves significant improvements compared to the state-of-the-art baselines. In this work, only character features and topic features are applied by our model. We consider applying more poetry-related features like rhyme, tones and genres on one side and on the other side to design more effective representations for these features to reinforce the attribution accuracy in the future.

Acknowledgments. The authors thank the anonymous reviewers for their insightful comments. This work is support in part by the National Natural Science Foundation of China (Grand No.61976124).

References

1. Sari, Y., Stevenson, M., Vlachos, A.: Topic or Style? exploring the most useful features for authorship attribution. In: Proceedings of the 27th International Conference on Computational Linguistics, pp. 343–353. Santa Fe, New Mexico (2018)
2. Chaski, C.E.: Who's at the keyboard? authorship attribution in digital evidence investigations. Int. J. Digit. Evid. **4**(1) (2005)
3. Abbasi, A., Chen, H.: Applying authorship analysis to extremist- group web forum messages. IEEE Intell. Syst. **20**(5), 67–75 (2005)
4. Frantzeskou, G., Stamatatos, E., Gritzalis, S., Katsikas, S.: Effective identification of source code authors using byte-level information. In: Proceedings of the 28th International Conference on Software Engineering, pp. 893–896. ACM Press, New York (2006)
5. Tianjiu, X., Ying, L.: Analysis of the words and n-grams in a dream of red mansions. New Technol. Library Inf. Serv. **257**(4), 50–57 (2015)
6. Mendenhall, T.C.: The characteristic curves of composition. Science (214S), 237- 246 (1887)
7. Stamatatos, E.: A survey of modern authorship attribution methods. J. Am. Soc. Inf. Sci. Technol. **60**(3), 538–556 (2008)
8. Hou, R., Huang, C.-R.: Robust stylometric analysis and author attribution based on tones and rimes. Natural Lang. Eng. 1–23 (2019)
9. Almodaresi, F., Ungar, L., Kulkarni, V., et al.: On the distribution of lexical features at multiple levels of analysis. In: Proceedings of the 55th Annual Meeting of the Association for Computational Linguistics (Short Papers), pp. 79–84. Vancouver, Canada (2017)
10. Bagnall, D.: Author identification using muti-headed recurrent neural networks. In: Working Notes Papers of the CLEF 2015 Evaluation Labs, volume 1391 (2015)
11. Shrestha, P., Sierra, S., Fabio, A., et al.: Convolutional neural networks for authorship attribution of short texts. In: Proceedings of the 15th Conference of the European Chapter of the Association for Computational Linguistics. vol. 2, pp. 669–674, Valencia, Spain, April 2017
12. Chen, D.: Identifying the authorship of the last forty chapters using mathematical linguistics: a discussion with Mr. Chen Bingzao. A Dream of Red Mansions, (5), 293–318 (1987)
13. Xiao, T., Liu, Y.: Analysis of Jin Yong and Gu Long's Novel Style Based on Clustering and Classification. J. Chin. Inf. Process. **29**(5), 167–17 (2015)
14. Nian, H., Chen, X., Wang, D.: Research on authorship attribution of contemporary literature. Comput. Eng. Appl. **46**(4), 226–229 (2010)
15. Wei, P.: From the distribution of common words examining the author issue of dream of red chamber author. In: Memorial Li Fanggui's 100th Anniversary International Symposium on Chinese History. Seattle: University of Washington (2002)
16. Jin, M., Jiang, M.: Text clustering on authorship attribution based on the features of punctuations usage. In: 2012 IEEE 11th International Conference on Signal Processing (ICSP), vol. 3. IEEE, pp. 2175–2178. Beijing, China (2012)
17. Jin, M.: Author identification based on n - gram pattern of auxiliary word. Measur. Lang. **23**(5), 225–240 (2002)
18. Ho, J.: From the use of three functional words "的", "地", "得" examining author's unique writing style–and on dream of red chamber author issues. BIBLID **120**(1), 119–150 (2015)
19. He, X., Liu, Y.: Mining stylistic features of rhythm and tempo base on text clustering. J. Chin. Inf. Process. **18**(6), 194–200 (2014)
20. Wu, C., Zhou, C.: Research on poetry style classification model based on frequent keyword co-occurrence. J. Xiamen Univ. (Natural Sci.) **47**(1), 41–44 (2008)
21. Yang, C., Sun, M., Yi, X., et al.: Stylistic Chinese Poetry Generation via Unsupervised Style Disentanglement EMNLP (2018)

22. Vaswani, A., et al.: Attention is all you need. In: Advances in Neural Information Processing Systems, pp. 6000–6010 (2017)
23. Blei, D.M., Ng, A.Y., Jordan, M.I.: Latent Dirichlet Allocation. J. Mach. Learn. Res. **3**, 993–1022 (2003)
24. Chang, J., Gerrish, S., Wang, C., et al.: Reading tea leaves: how humans interpret topic models. Adv. Neural Inf. Process. Syst. pp. 288–296 (2009)
25. David Mimno, Hanna Wallach, Edmund Talley et al.: Optimizing semantic coherence in topic models. In: Proceedings of the 2011 Conference on Emperical Methods in Natural Language Processing, pp. 262–272. Edinburgh, Scotland, UK. Association of Computational Linguistics (2011)
26. Newman, D., Noh, Y., Edmund, T., et al.: Evaluating topic models for digital libraries. In: Proceedings of the 10th Annual Joint Conference on Digital Libraries, JCDL 2010, pp. 215–224. ACM, New York, NY, USA (2010)
27. Bahdanau, D., Cho, K., Bengio, Y.: Neural Machine Translationby Jointly Learning to Align and Translate ICLR (2015)
28. Yi, Y., Zheng, Y., He, Z.: Discrimination of Classical Poetry Authors Based on Machine Learning. Mind and calculation, (03) (2007)
29. Devlin, J., Chang, M.W., Lee, K., Toutanova, K.: BERT: pre-trainingof deep bidirectional transformers for language understanding. In: Proceedingsof the 2019 Conference of the North American Chapter of the Association for Computational Linguistics: Human Language Technologies, Volume 1 (Long and Short Papers), pp. 4171–4186 (2019)

Aspect Fusion Graph Convolutional Networks for Aspect-Based Sentiment Analysis

Fuyao Zhang, Yijia Zhang, Shuo Hou, Fei Chen, and Mingyu Lu[✉]

School of Information Science and Technology, Dalian Maritime University, Dalian 116024,
Liaoning, China
lumingyu@dlmu.edu.cn

Abstract. Aspect-based sentiment classification aims to distinguish the sentiment
polarities over aspect terms in a sentence. Recent approaches to aspect-based sentiment classification use graph-based models to integrate the syntactic structure
of sentences. While being practical, these methods ignore the close relationship
between the topological structure of the dependency tree and the dependency distance. To solve this problem, we propose to build an Aspect Fusion Graph Convolutional Network (AFGCN) of sentences to take advantage of syntactic information and word dependencies. Specifically, we enhance the syntactic dependencies
of each instance by introducing dependency tree and dependency-position graph.
Then, we use two graph convolutional networks to fuse the dependency tree and the
dependency-position graph to generate the interactive emotion features of aspects.
Finally, we use a novel attention mechanism to fully integrate the significant features related to aspect semantics in the hidden state vectors of the convolution
layer and the masking layer. Extensive experiments on five benchmark datasets
show that our method achieves state-of-the-art performance.

Keywords: Aspect-based sentiment analysis · Graph convolutional network ·
Syntactic dependency

1 Introduction

Aspect-based sentiment analysis (ABSA) aims at fine-grained sentiment analysis of
sentiment texts such as product reviews. More specifically, ABSA involves two tasks:
(1) identifying various aspects of a sentence, (2) determining the sentiment polarity (for
example, positive, negative, neutral) expressed in a particular aspect. This paper focuses
on the second task: Aspect-based Sentiment Classification. For example, in a comment
about a laptop saying, "From the speed to the multitouch gestures this operating system
beats Windows easily.", the sentiment polarities for two aspects of operating system and
Windows are positive and negative, respectively. In the task of aspect sentiment analysis,
we need to distinguish sentiment polarity according to different aspects.

In the early research of ABSA (Jiang et al., Mohammad et al.) [1, 2], machine
learning algorithm is often used to construct sentiment classifiers. Dependency-based
parse trees are used to provide more comprehensive syntax information. Therefore, the

© Springer Nature Switzerland AG 2021
H. Lin et al. (Eds.): CCIR 2021, LNCS 13026, pp. 74–87, 2021.
https://doi.org/10.1007/978-3-030-88189-4_6

whole dependency tree can be encoded from leaf to root by recursive neural network (RNN) (Dong et al., Nguyen et al., Wang et al.) [3–5]. Then various neural network models (Dong et al., Vo et al., Chen et al.) [3, 6, 7] are proposed, including long short-term memory network (LSTM) (Wang et al.) [8], convolutional neural network-based (CNN) (Huang et al., Li et al.) [9, 10], and memory-based (Tang et al.) [11] or hybrid methods (Xue et al.) [12], or the distance of the internal node can be calculated and used for attention weight decay (He et al.) [13]. These models represent a sentence as a word sequence, ignoring the syntactic relationship between words, making it difficult for them to find words far away from the expected words. In recent years, several studies have used graph-based models to combine sentence syntactic structure (Zhang et al., Sun et al., Huang et al., Liang et al., Chen et al.) [9, 14–17], which has better performance than the model without considering syntactic relationships. However, the above model only fully considers the topology structure of dependency tree, or the actual distance between words, but does not fully play to the advantages of dependency tree, and does not fully integrate the topology structure of dependency tree and dependency distance. The shortcomings of these approaches should not be overlooked.

To better capture opinion features for aspect sentiment classification, we propose the AFGCN model, which fully combines the topological structure and the dependency distance calculated from the dependency tree. Inspired by the position mechanism [18], this model aggregates valid features in an LSTM-based architecture and uses the syntactic proximity of a context word to the aspect, also known as proximity weight, to determine its importance in a sentence. At the same time, we apply GCN network on dependency tree and dependency-position graph, respectively. We can use long-range multiword relations and syntactic information through GCN to potentially draw syntactically related words to the target. The output is fed into a masking mechanism, which filters out non-aspect words to get focused aspect features. Aspect-specific features are fed into the LSTM output, and the aspect fusion attention mechanism is used to update the most relevant features. After all operations above, the representation of context and aspects concentrated on passing through a linear layer to get the final output. Experiments demonstrate the effectiveness of our model. The main contributions of this paper are presented as follows:

- We build a complex task-specific syntactic dependency module, which profoundly integrates dependency tree and dependency-position graph to enhance the syntactic dependency of each instance.
- An aspect fusion graph convolutional network model (AFGCN) was proposed, which combined attention mechanism to fully integrate prominent features related to aspect semantics in the hidden state vectors of the convolutional layer and the masking layer, to fully combine the topology structure and dependency distance of the dependency tree.
- Experimental results on five benchmark datasets show the effectiveness of our proposed model in capturing them in aspect-based sentiment classification.

2 Related Work

In aspect sentiment analysis, some early work focused on using machine learning algorithms to capture sentiment polarity based on rich features of content and syntactic structure (Jiang et al., Kiritchenko et al.) [1, 2]. The latest development of aspect-level sentiment classification (ASC) focuses on developing various types of deep learning models. The neural models without considering the syntactic models can be divided into several types: LSTM based (Tang et al., 2016a; Ma et al., 2017) [11, 19], CNN based (Huang et al., Li et al.) [9, 10], memory-based methods (Tang et al., Chen et al.) [7, 11], etc. In neural network approaches, some use RNN variants (such as LSTM and GRU) to model the sentence representation (Majumder et al.) [20].

Syntactic information allows dependency information to be kept in long sentences and helps to bridge the gap between aspects and opinion words. Tai et al. [21] proposed a tree-structured LSTM, which enables people to learn the dependency information between words and phrases. Mouetal et al. [22] utilizes the short path of dependency trees and uses convolutional neural networks to learn the representation of sentences. Recently, some studies use graph-based models to integrate syntactic structures. Zhang et al. [14] uses GCN to capture specific aspects of syntactic information and word dependency on the syntactic dependence tree. Liang et al. [23] proposes an Interactive Graph Convolutional Networks (InterGCN) model to extract both aspect-focused and inter-aspects sentiment features for the specific aspect. Zhang et al. [24] convolutes over hierarchical syntactic and lexical graphs and builds a concept hierarchy on both the syntactic and lexical graphs for differentiating dependency relations.

These observations enable us to build a neural model of dependency trees that fully integrates syntactic dependence and distance and makes accurate sentiment predictions about certain aspects. Specifically, we propose an Aspect Fusion Graph Convolutional Networks model (AFGCN).

3 The Proposed Model

The overall architecture of the proposed AFGCN model is shown in Fig. 1. We first assume a sentence with n words and m aspects from the SemEval-2014 dataset, i.e. $s = \{w_0, w_1, ..., w_a, w_{a+1}, ..., w_{a+m-1}, ..., w_{n-1}\}$, where w_i represents the i -th contextual word and w_a represents the start token of aspect words. Each word is embedded into a low-dimensional real-valued vector with a matrix $V \in R^{|N| \times d_i}$, where $|N|$ is the size of the dictionary while d_i is the dimension of a word vector. We use the pre-trained word embedding GloVe to initialize the word vectors, and the resulting word embeddings are adopted to a bidirectional LSTM to produce the sentence hidden state vectors h_t. Since the input representation already contains aspect information, the context representation specific to the aspect is obtained by linking the hidden state from both directions: $h_t = [\overrightarrow{h_t} ; \overleftarrow{h_t}]$ where $\overrightarrow{h_t}$ is the hidden state from the forward LSTM and $\overleftarrow{h_t}$ is from the backward.

3.1 Producing Dependency Tree

We use spacy[1] to construct a given sentence into a directed dependency tree.

Then we construct the adjacency matrix based on the directed dependency tree, and we set all the diagonal elements of the matrix to 1. If there is a dependency between two words, we also write down the corresponding position in the matrix as 1.

And then, an adjacency matrix $M_{ij}^T \in R^{n \times n}$ is derived from the dependency tree of the input sentence.

3.2 Producing Dependency-Position Graph

To highlight the relationship between context and aspect, we compute the relative position weight of each element of the adjacency matrix according to aspect.

$$W_{i,j}^F = \begin{cases} 1 & \text{if } w_i \in \{a_i^s\} \text{ and } w_j \in \{a_i^s\} \\ 1/(|j - p^b| + 1) & \text{if } w_i \in \{a_i^s\} \\ 1/(|i - p^b| + 1) & \text{if } w_j \in \{a_i^s\} \\ 0 & \text{otherwise} \end{cases} \tag{1}$$

where $|\cdot|$ is an absolute value function, p^b is the beginning position of the aspect, $\{a^s\}$ is the word set of the aspect.

To establish a closer dependency relationship between context words, we integrate ordinary dependency graph $D_{i,j}^G$, which is obtained by the adjacency matrix of the dependency tree symmetrically along the diagonal, and relative position weight $W_{i,j}^G$ to derive the adjacency matrix of the dependency-position graph.

$$M_{i,j}^G = \begin{cases} 1 + W_{i,j}^G & \text{if } D_{i,j}^G = 1 \\ W_{i,j}^G & \text{otherwise} \end{cases} \tag{2}$$

3.3 Proximity-Weight Convolution

Previous dependency tree-based models mainly focus on the topology of the dependency tree or the distance of the dependency tree. However, few models apply them together, limiting the effectiveness of these models in identifying key context words used in representation. This syntactic dependency information is formalized as an adjacent weight in our proposed model, which describes the proximity between context and aspect. Recall the example of the dependency tree in Fig. 1: "But the staff was so horrible to us." The actual distance between the aspect "staff" and the sentiment word "horrible" is 3, but the dependency distance is 1. Intuitively, dependency distance is more beneficial to aspect-based sentiment classification than ordinary distance.

---

[1] In this work, we use spaCy toolkit for producing the dependency tree of the input sentence: https://spacy.io/.

We construct a dependency tree and then compute the dependency distance for the context words: the length of the shortest dependency path between the aspect and the sentiment words. If the aspect contains multiple words, we minimize the dependency distance between the context and all aspect words. The dependency proximity weights of the sentence are computed by the formula below:

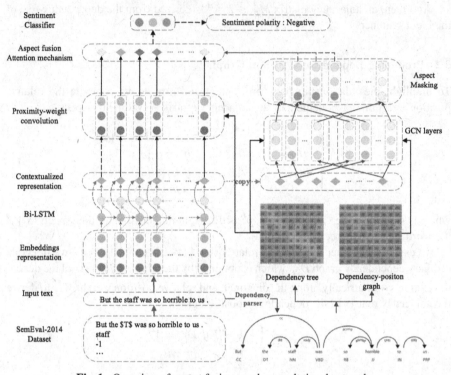

Fig. 1. Overview of aspect fusion graph convolutional network.

$$p_i = \begin{cases} 1 - \frac{d_i}{n} & 0 \le i < \tau \, or \, \tau + m \le i < n \\ 0 & \tau \le i < \tau + m \end{cases} \tag{3}$$

where proximity weight $p_i \in R$, d_i is the dependency distance from the word to aspect in the sentence.

Inspired by Zhang et al. [14], we introduce proximity-weight convolution. Unlike the original definition of convolution, proximity-weighted convolution allocates proximity weights before convolution calculation. It is essentially a one-dimensional convolution with a length-l kernel. The proximity-weight convolution process is then assigned as:

$$q_i \in max\left(W_c^T [r_{i-t} \oplus \cdots \oplus r_i \oplus \cdots \oplus r_{i+t}] + b_c, 0 \right) \tag{4}$$

where $r_i = p_i h_i$ and $t = \lfloor \frac{l}{2} \rfloor$, $r_i \in R^{2d_h}$ represents the proximity-weighted representation of the i-th word in the sentence, $q_i \in R^{2d_h}$ represents the feature representation obtained from the convolution layer, and $W_c \in R^{l \cdot 2d_h \times 2d_h}$ and $b_c \in R^{2d_h}$ are weight and bias of the convolution kernel, respectively.

3.4 Aspect Fusion Graph Convolutional Network

Aiming to take advantage of syntactic dependency, we use two graph convolutional networks to fuse dependency tree and dependency-position graph, respectively, to generate interactive sentiment features for aspect. The representation of each node is calculated with graph convolution with normalization factor, and the representation of each node is updated according to the hidden representations of its neighborhood:

$$h_i^l = Relu((\sum_{j=1}^{n} M_{ij} W^l g_j^{l-1})/(d_i + 1) + b^l) \tag{5}$$

$$g_j^l = P(h_i^l) \tag{6}$$

where $g_j^{l-1} \in R^{2d_h}$ is the representation of the j-th token evolved from the preceding GCN layer. $P(\cdot)$ is a PairNorm function that integrates position-aware transformation and has been used in previous GCN network (Xu et al., Zhao et al.) [25, 26]. M_{ij} includes M^G and M^T, we take these two matrices integrating different dependency relationships as the inputs of two groups of GCN, respectively. D_i is the degree of the i-th token in the tree. W^l and b^l are trainable parameters, respectively.

Then, we can capture the final representation of the GCN layers from different inputs, h^G and h^T, where h^G is the representation of M^G and h^T is the representation of M^T. And thus, inspired by Liang et al. [23], we combine these two final representations to extract the interactive relations between dependency-position feature and dependency feature:

$$\tilde{h} = h^G + \gamma h^T \tag{7}$$

where γ is the coefficient of dependency feature. The combination method takes into account both syntactical dependence and long-term multi-word relations. We use aspect masking to mask non-aspect representations to highlight the critical features of aspect words. In other words, we keep the final representation of the aspect words output by the GCN layer and set the final representation of the non-aspect words to 0.

3.5 Aspect Fusion Attention Mechanism

We intend to fuse the significant features related to aspect semantic in the hidden state vectors of the convolutional layer and the masking layer through a new way of Aspect fusion Attention mechanism, and setting accurate attention weight for each contextual word accordingly. The attention weights assigning process is formulated below:

$$u_{it} = tanh(W_w q_i h_i^M + b_w) \tag{8}$$

$$\alpha_t = \frac{exp(u_{it}^T u_w)}{\sum_{i=1}^{n} exp(u_{it}^T u_w)} \tag{9}$$

where h_i^M and q_i are the final hidden state vectors output by the Masking layer and the convolution layer respectively. W_w and U_w are weights that are randomly initialized. Then we use the formula $r = \sum_{t=1}^{n} \alpha_t q_i$ to get the corresponding attention weight.

Table 1. Statistics of the experimental datasets.

Dataset	Positive		Neural		Negative	
	Train	Test	Train	Test	Train	Test
Twitter	1561	173	3127	346	1560	173
Lap14	994	341	464	169	870	128
Rest14	2164	728	637	196	807	196
Rest15	912	326	36	34	265	182
Rest16	1240	469	69	30	439	117

3.6 Model Training

The aspect-based representation r is passed to a fully connected softmax layer whose output is a probability distribution over the different sentiment polarities.

$$p = \text{softmax}(W_p r + b_p) \tag{10}$$

where W_p and b_p are learnable parameters for the sentiment classifier layer.

The model is trained with the standard gradient descent algorithm by minimizing the cross-entropy loss on all training samples:

$$\zeta = -\sum_i^J p_i log\ \hat{p}_i + \lambda ||\Theta|| \tag{11}$$

where J is the number of training samples, p_i and $\hat{p}_i$ is the ground truth and predicted label for the i-th sample, Θ represents all trainable parameters, and λ is the coefficient of L2-regularization.

4 Experiments

4.1 Datasets and Experimental Settings

Datasets

Our experiments are conducted on five datasets: one is the twitter benchmark dataset constructed by dong et al. [3]. The other four are from the SemEval2014 (Pontiki et al., 2014) [27], SemEval2015 (Pontiki et al., 2015) [26], and SemEval2016 (Pontiki et al., 2016) [28] benchmark datasets, which are composed of two types of data: laptops and restaurants. Each sample is composed of comment sentences, aspects, and the sentiment polarity of the aspects. Building on previous work, we remove samples with conflicting polarity and undefined aspects in rest15 and rest16 sentences. The statistics for the datasets are shown in Table 1.

Settings

For the fairness of model comparison, we use similar parameters in the comparison model. In all experiments, we use 300-dimensional preprocessing GloVe vectors (Pennington et al.) [32] as initial word embeddings. The dimension of the hidden state vector is set to 300. To train the model, we use Adam as the optimizer with a learning rate of 0.001. The coefficient of L2-regularization is 10^{-5}, the coefficient γ is set to 0.2, and the batch size is 32. Besides, the number of GCN layers is set to 2, which is the best performing depth in the pilot study. We adopt Accuracy and Macro-Averaged F1 as the evaluation metrics.

Table 2. Comparison results for all methods in terms of accuracy and F1 (%). The best results on each dataset are in bold.

Model	Twitter		Lap14		Rest14		Rest15		Rest16	
	Acc.	F1	Acc.	F1	Acc.	F1	Acc.	F1	Acc.	F1
SVM	63.40	63.30	70.49	–	80.16	–	–	–	–	–
ATAE-LSTM	69.65	67.40	69.14	63.18	77.32	66.57	75.43	56.34	83.25	63.85
Mem-Net	71.48	69.90	70.64	65.17	79.61	69.64	77.31	58.28	85.44	65.99
RAM	69.36	67.30	74.49	71.35	80.23	70.80	79.30	60.49	85.58	65.76
TNet-LF	72.98	71.43	74.61	70.14	80.42	71.03	78.47	59.47	89.07	70.43
TD-GAT	72.20	70.45	75.63	70.74	81.32	71.72	80.38	60.50	87.71	67.87
ASGCN	72.15	70.40	75.55	71.05	80.77	72.02	79.89	61.89	88.99	67.48
kumaGCN	72.45	70.07	76.12	72.42	81.43	73.64	80.69	65.99	89.39	**73.19**
BiGCN	74.16	**73.35**	74.59	71.84	81.97	73.48	**81.16**	64.79	88.96	70.84
AFGCN	**74.69**	73.23	**77.43**	**73.64**	**82.50**	**73.66**	79.89	**66.29**	**89.61**	72.02

4.2 Models for Comparison

A comprehensive comparison is carried out between our proposed model (AFGCN) against several state-of-the-art baseline models, as listed below:

- SVM (Kiritchenko et al.) [2] is the model which has won SemEval 2014 task 4 with conventional feature extraction methods.
- ATAE-LSTM (Wang et al.) [8] is a classic LSTM based model which explores the relationship between aspect and the content of an attention-based LSTM sentence.
- Mem-Net: (Tang et al.) [11] utilizes multi-hops attention to the context words used for sentence representation to illustrate the importance of each context word.

- RAM (Chen et al.) [7] uses multi-hops of attention layers and combines the outputs with a RNN for sentence representation.
- TNet-LF (Li et al.) [10] puts forward Context-Preserving Transformation (CPT) to preserve and strengthen the informative part of contexts.
- TD-GAT (Huang et al.) [9] proposes a graph attention network to explicitly utilize the dependency relationship among words.
- ASGCN (Zhang et al.) [13] employs a GCN over the dependency tree to exploit syntactical information and word dependencies.
- BiGCN (Zhang et al.) [23] convolutes over hierarchical syntactic and lexical graphs and build a concept hierarchy on both the syntactic and lexical graphs.
- kumaGCN (Chen et al.) [16] propose gating mechanisms to dynamically combine information from word dependency graphs and latent graphs.

Among the baselines, the first five methods are classic models with typical neural structures. The bottom five methods are graph-based and syntax-integrated ones.

We reproduce the results for baselines if the authors provide the source code. For the methods (TD-GAT) with no released code, we implement them by ourselves using the optimal hyperparameters settings reported in their papers. In our experiments, since we report the results over three runs with the random initialization, we stop training when the F1 score does not increase for a certain number (5) of rounds at one run.

4.3   Performance Comparison

The comparison results for all methods are shown in Table 2. From these results, we make the following observations.

Our proposed model AFGCN shows significant improvements on the five datasets. Table 2. shows the performance comparisons. Our method outperforms SVM by 2.34 and 6.94 Acc. score on Rest14 and Lap14, respectively. This indicates that our neural approach extracts more practical features than hardcoded feature engineering. Our model achieves the best performance on three datasets (Lap14, Rest14 and, Rest15) and is only 0.12% lower than the F1. score of the best-performing model on the Twitter dataset. However, our model performed poorly on the Rest16 dataset, with a 1.17% difference in F1. score from the best-performing kumaGCN. We speculate that the reason for its poor performance may be caused by the different distribution of positive, neutral and negative sentiment between the train set and the test set, as shown in Table 1.

The methods based on the combination of graph and syntax (TD-GAT, ASGCN, kumaGCN, and BiGCN) are significantly better than the first five methods without considering syntax, indicating that the dependency relationship is beneficial to the recognition of sentiment polarity, which is consistent with previous studies. However, they are worse than the AFGCN model we proposed because our proposed model fully integrates topology structure and dependent distance. The result proves that our AFGCN model, which combines dependency tree and dependency distance, is helpful to improve performance.

The large performance gaps between our model and baseline models confirm the effectiveness of our proposed architecture. We believe that using context and dependency information from the sentence, we can encode aspect vectors through proximity-weight convolution and GCN layers. Proximity-weight convolution and GCN layers can be considered messaging networks that propagate information along word sequence chains or syntactic dependency paths. Since relevant information is transmitted to aspect, we only need a simple Attention mechanism to encode the weighted information in significant words, thus preserving information relevant to the categorization task.

4.4 Ablation Study

We conduct an ablation study further to analyze the impact of different components of AFGCN. The results are shown in Table 3.

Table 3. Ablation study results (%). Acc. represents accuracy, F1 represents Macro-F1 score.

Model	Twitter		Lap14		Rest14		Rest15		Rest16	
	Acc.	F1	Acc.	F1	Acc.	F1	Acc.	F1	Acc.	F1
AFGCN w/o P-w	72.98	71.46	**77.58**	**73.70**	80.71	72.01	79.73	63.35	89.11	69.18
AFGCN w/o GCN	73.12	70.89	75.70	71.96	81.33	73.33	78.22	60.67	88.96	71.26
AFGCN w/o Att	73.84	72.19	76.95	72.84	80.62	72.38	79.33	63.28	88.79	**73.24**
AFGCN w/o tree	73.98	72.28	76.95	73.32	81.25	72.04	80.07	65.08	88.96	71.19
AFGCN w/o graph	73.55	71.98	75.70	71.82	82.31	73.23	79.52	64.35	88.14	71.15
AFGCN	**74.69**	**73.23**	77.43	73.64	**82.50**	**73.56**	**79.89**	**66.29**	**89.61**	72.02

First, removal of proximity-weight convolution (i.e., AFGCN w/o P-w.) degrades the performance of four datasets but improves the performance for about 0.1% of Lap14 datasets. We argue that if the syntax is not essential to the data, then the integration of adjacent weights does not help reduce the noise of user-generated content.

Second, the removal of GCN layers is generally an evident performance degradation. Thus, it can be seen that GCN layers promote the development of AFGCN to a great extent because GCN captures both syntactic lexical dependencies and long-range lexical relationships. We can also observe that removing "aspect fusion Attention mechanism" (i.e., AFGCN w/o Att.) slightly degrades performance, indicating that our Attention mechanism helps integrate significant features related to aspect semantics in sentences and is an integral part of AFGCN.

Then we investigate the impacts of dependency tree (i.e., AFGCN w/o tree.) and dependency-position graph (i.e., AFGCN w/o graph.) Compared with the complete AFGCN, the performance of both is degraded, indicating that the effect of one graph (tree) is not as good as that of two fused graphs. We also found that the two compete on Rest datasets, with each having their contribution from a lexical and syntactic perspective.

4.5 Impact of GCN Layers

We investigate the effect of the number of layers on the performance of our proposed AFGCN. We vary the layer number from 1 to 8 and report the results in Fig. 2. It can be seen that our model achieves the best results with two layers, and thus we set the number of GCN layers as 2 in our experiments. Using only one layer of AFGCN is not enough to obtain specific syntactic dependencies of the context on aspect. However, the performance does not constantly get improved with the increasing number of layers. The performance of AFGCN fluctuates with the increase of the number of GCN layers and basically decreases when the model depth is greater than 2. Analysis implies that a larger model introduces more parameters, resulting in a less generic model and challenging to train.

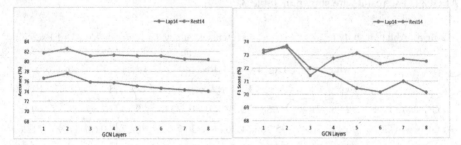

Fig. 2. Impacts of GCN layers.

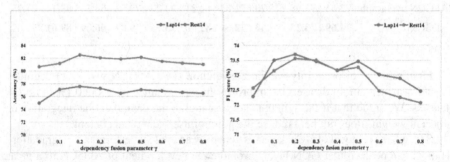

Fig. 3. Impacts of the dependency fusion parameter γ.

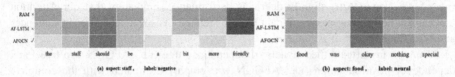

Fig. 4. Visualization results for RAM, AF-LSTM and AFGCN, where $\checkmark$ and $\times$ denotes the correct and wrong prediction, respectively.

4.6 Impact of the Dependency Fusion Parameter γ

To investigate how the trade-off between using dependency-position graph and dependency tree affects AFGCN performance, we use a step size of 0.1 to vary γ from 0 to 0.8.

Figure 3 shows the F1. scores obtained by AFGCN on Lap14 and Rest14 with different γ. When $\gamma = 0$, the model degenerates to GCN of fused dependency-position graph only. It can be observed that the performance significantly improves with the increase of γ value from 0 to 0.2, indicating that the fusion of graph and tree is beneficial to focusing the aspect related features. the curve reaches its maximum value when $\gamma = 0.2$, indicating that the dependency-position graph and dependency tree structure in this model are complementary. When γ is greater than 0.2, the curve shows a fluctuating downward trend. Thus, we set $\gamma = 0.2$.

4.7 Case Study and Error Analysis

To gain more insights into our model's behavior, we show two case studies in Fig. 4. We visualize the attention scores, the predicted and the ground truth labels for these examples.

As can be seen from Fig. 4(a), the aspect for the given example is "staff" with negative sentiment, and only our model predicts correct sentiment. This example uses the subjunctive word "should", which makes it extra difficult to detect grammar. Due to the lack of syntax information, RAM and AF-LSTM cannot make the right decision for the two examples. Both models assign the highest weight to the word "friendly", which is an irrelevant sentiment word to this target, leading to an incorrect prediction. In contrast, our model assigns the largest weight to the sentiment keyword "should" and correctly predict the negative polarity of the aspect "staff" in the first sentence.

Figure 4(b) shows the examples for error analysis. RAM gives relatively high attention weight to the words "nothing" and "special", but it still predicts the wrong sentiment polarity. Although AF-LSTM calculates the relationship between the context and the aspect, the short distance between "food" and "okay" causes the LSTM to assign the most significant attention scores to "okay". On the other hand, since "good" and "food" are closely related in the dependency tree, the solid positive polarity of "good" also prejudices the AFGCN decision. This type of error frequently appears in neutral cases. If negative expression (e.g., "nothing", "shouldn't") is related to aspect, the neural model does not differentiate well.

5 Conclusion

Previous methods for aspect-based sentiment classification depended on the syntactic relationship between aspect and context often ignore the dependency distance relationship between context. In this paper, we have built a framework that leverages graph-based approaches and syntactic dependencies between contextual terms and aspect to construct an applicable model. In addition to dependency tree, we built a dependency-position graph to enhance the syntactic dependencies of each instance. And we propose an aspect fusion graph convolutional network model to fully combine the topology structure and dependency distance of dependency tree. Finally, we design the aspect fusion attention module to fully integrate the significant features related to aspect semantics in the hidden state vectors of the convolution layer and the masking layer. Experimental results demonstrate the effectiveness of our proposed model and suggest that dependency distance and syntactic dependency are more beneficial to aspect-based sentiment classification.

Acknowledgment. This work is supported by grant from the Natural Science Foundation of China (No. 61976124) and Social and Science Foundation of Liaoning Province (No. L20BTQ008).

References

1. Jiang, L., Yu, M., Zhou, M., Liu, X., Zhao, T.: Target-dependent twitter sentiment classification. In: Proceedings of the 49th Annual Meeting of the Association for Computational Linguistics: Human Language Technologies, pp. 151–160 (2011)
2. Kiritchenko, S., Zhu, X., Cherry, C., Mohammad, S.: NRC-Canada-2014: detecting aspects and sentiment in customer reviews. In: Proceedings of the 8th International Workshop on Semantic Evaluation (SemEval 2014), pp. 437–442 (2014)
3. Dong, L., Wei, F., Tan, C., Tang, D., Zhou, M., Xu, K.: Adaptive recursive neural network for target-dependent twitter sentiment classification. In: Proceedings of the 52nd Annual Meeting of the Association for Computational Linguistics (Volume 2: Short Papers), pp. 49–54 (2014)
4. Nguyen, T.H., Shirai, K.: Phrase RNN: phrase recursive neural network for aspect-based sentiment analysis. In: Proceedings of the 2015 Conference on Empirical Methods in Natural Language Processing, pp. 2509–2514 (2015)
5. Wang, W., Pan, S.J., Dahlmeier, D., Xiao, X.: Recursive neural conditional random fields for aspect-based sentiment analysis. arXiv preprint arXiv:1603.06679 (2016)
6. Vo, D.T., Zhang, Y.: Target-dependent twitter sentiment classification with rich automatic features. In: Twenty-Fourth International Joint Conference on Artificial Intelligence (2015)
7. Chen, P., Sun, Z., Bing, L., Yang, W.: Recurrent attention network on memory for aspect sentiment analysis. In: Proceedings of the 2017 Conference on Empirical Methods in Natural Language Processing, pp. 452–461 (2017)
8. Wang, Y., Huang, M., Zhu, X., Zhao, L.: Attention-based LSTM for aspect-level sentiment classification. In: Proceedings of the 2016 Conference on Empirical Methods in Natural Language Processing, pp. 606–615 (2016)
9. Huang, B., Carley, K.M.: Parameterized convolutional neural networks for aspect level sentiment classification. arXiv preprint arXiv:1909.06276 (2019)
10. Xin, L., Bing, L., Lam, W., Bei, S.: Transformation networks for target-oriented sentiment classification. In: Proceedings of the 56th Annual Meeting of the Association for Computational Linguistics (Volume 1: Long Papers) (2018)
11. Tang, D., Qin, B., Feng, X., Liu, T.: Effective LSTMs for target-dependent sentiment classification. arXiv preprint arXiv:1512.01100 (2015)
12. Xue, W., Li, T.: Aspect based sentiment analysis with gated convolutional networks. arXiv preprint arXiv:1805.07043 (2018)
13. He, R., Lee, W.S., Ng, H.T., Dahlmeier, D.: Effective attention modeling for aspect-level sentiment classification. In: Proceedings of the 27th International Conference on Computational Linguistics, pp. 1121–1131 (2018)
14. Zhang, C., Li, Q., Song, D.: Aspect-based sentiment classification with aspect-specific graph convolutional networks. arXiv preprint arXiv:1909.03477 (2019)
15. Sun, K., Zhang, R., Mensah, S., Mao, Y., Liu, X.: Aspect-level sentiment analysis via convolution over dependency tree. In: Proceedings of the 2019 Conference on Empirical Methods in Natural Language Processing and the 9th International Joint Conference on Natural Language Processing (EMNLP-IJCNLP), pp. 5683–5692 (2019)

16. Liang, B., Du, J., Xu, R., Li, B., Huang, H.: Context-aware embedding for targeted aspect-based sentiment analysis. arXiv preprint arXiv:1906.06945 (2019)

17. Chen, C., Teng, Z., Zhang, Y.: Inducing target-specific latent structures for aspect sentiment classification. In: Proceedings of the 2020 Conference on Empirical Methods in Natural Language Processing (EMNLP), pp. 5596–5607 (2020)

18. Fan, C., Gao, Q., Du, J., Gui, L., Xu, R., Wong, K.: Convolution-based memory network for aspect-based sentiment analysis. The 41st International ACM SIGIR Conference on Research Development in Information Retrieval (2018)

19. Ma, D., Li, S., Zhang, X., Wang, H.: Interactive attention networks for aspect-level sentiment classification. arXiv preprint arXiv:1709.00893 (2017)

20. Majumder, N., Poria, S., Gelbukh, A., Akhtar, M.S., Cambria, E., Ekbal, A.: IARM: inter-aspect relation modeling with memory networks in aspect-based sentiment analysis. In: Proceedings of the 2018 Conference on Empirical Methods in Natural Language Processing, pp. 3402–3411 (2018)

21. Tai, K.S., Socher, R., Manning, C.D.: Improved semantic representations from tree-structured long short-term memory networks. arXiv preprint arXiv:1503.00075 (2015)

22. Liang, B., Yin, R., Gui, L., Du, J., Xu, R.: Jointly learning aspect-focused and inter-aspect relations with graph convolutional networks for aspect sentiment analysis. In: Proceedings of the 28th International Conference on Computational Linguistics, pp. 150–161 (2020)

23. Zhang, M., Qian, T.: Convolution over hierarchical syntactic and lexical graphs for aspect level sentiment analysis. In: Proceedings of the 2020 Conference on Empirical Methods in Natural Language Processing (EMNLP), pp. 3540–3549 (2020)

24. Xu, L., Bing, L., Lu, W., Huang, F.: Aspect sentiment classification with aspect-specific opinion spans. arXiv preprint arXiv:2010.02696 (2020)

25. Zhao, L., Akoglu, L.: PairNorm: tackling oversmoothing in GNNs. arXiv preprint arXiv: 1909.12223 (2019)

26. Pontiki, M., Galanis, D., Pavlopoulos, J., Papageorgiou, H., Androutsopoulos, I., Manandhar, S.: Semeval-2014 task 4: aspect based sentiment analysis. In: COLING 2014 (2014)

27. Pontiki, M., Galanis, D., Papageorgiou, H., Manandhar, S., Androutsopoulos, I.: Semeval-2015 task 12: aspect based sentiment analysis. In: Proceedings of the 9th International Workshop on Semantic Evaluation (SemEval2015), pp. 486–495 (2015)

28. Pennington, J., Socher, R., Manning, C.D.: GloVe: global vectors for word representation. In: Proceedings of the 2014 Conference on Empirical Methods in Natural Language Processing (EMNLP), pp. 1532–1543 (2014)

Iterative Strict Density-Based Clustering for News Stream

Kaijie Shi[1,2], Jiaxin Shi[1], Yu Zhou[1], Lei Hou[1,3](✉), and Juanzi Li[1,3]

[1] Department of Computer Science and Technology, BNRist, Tsinghua University,
Beijing, China
`skj19@mails.tsinghua.edu.cn, bryanzhou008@ucla.edu,`
`{houlei,lijuanzi}@tsinghua.edu.cn`
[2] Tsinghua Shenzhen International Graduate School, Tsinghua University,
Shenzhen, China
[3] KIRC, Institute for Artificial Intelligence, Tsinghua University,
Beijing, China

Abstract. News Streams are booming with the prosperity of the Internet, leading to increased demand for an efficient and effective news clustering method. Since news reports vary greatly in different countries, languages and news-topics, clustering diverse news has proven to be a big challenge for all researchers. The results of current clustering methods expose their inability to detect fine-grained topics. They tend to detect topics on a coarse-grained scale, resulting in clustering different fine-grained topics together.

In this paper, we propose Iterative Strict Density-based Clustering (ISDC), a new approach for detecting fine-grained topics in an evolving news stream. The main idea of ISDC is to keep every cluster as a high-density cluster throughout the news stream by iteratively splitting growing clusters. We further apply multilingual-sentence-bert instead of word embedding as the news encoder to improve the news representation quality. We conduct comprehensive experiments on two datasets and demonstrate the superiority of our proposed method.

Keywords: Streaming clustering · Iterative density-based clustering · Fine-grained topic detection

1 Introduction

Topic Detection and Tracking (TDT) [3] is an information processing technology designed to help people cope with the increasingly serious Internet information explosion problem. It aims to automatically identify new topics and keep track of known topics in the information flow of news media. As a key link of TDT, stream clustering aims to find news topics in evolving data streams in one pass using a limited amount of memory.

Density-based algorithms are an important group of stream clustering. By adopting the online-offline paradigm [19] with micro-clusters which are defined as

© Springer Nature Switzerland AG 2021
H. Lin et al. (Eds.): CCIR 2021, LNCS 13026, pp. 88–99, 2021.
https://doi.org/10.1007/978-3-030-88189-4_7

high-density clusters, density-based algorithms consist of two main steps: Firstly all samples are assigned to different micro-clusters online, then these micro-clusters are merged into final clusters through offline density clustering. However, density-based algorithms are not satisfactory in terms of accuracy because of their loose restriction. In the online period, since the algorithm only compares samples with existing micro-clusters centers, news of the same cluster may be different. In addition, the offline clustering step only loosely limits the distance between different micro-cluster centers. As a result, the differences of samples within one micro-cluster becomes transitive, which further reduces cohesion in the final clustering result. Therefore, a strict restriction to all samples is indeed necessary for high-quality clustering in fine-grained topic detection.

To alleviate these problems of density-based algorithms, we propose Iterative Strict Density-based Clustering (ISDC). ISDC maintains a cluster set which only contains *topic clusters*. A *topic cluster* is a cluster where the distance between each sample is less than a certain value. When a new samples arrives, we try to insert it into the nearest cluster. After this insertion, if the corresponding cluster is no longer a topic cluster, ISDC will split the cluster iteratively using DBSCAN [8] until all sub-clusters become topic clusters. Furthermore, ISDC will gradually decrease the time weight of outdated clusters. Through sequential updates and iterative spliting, we group similar samples together and keep dissimilar samples far apart. Compared with other algorithms, ISDC can better distinguish different topics while maintaining computational equilibrium. Experimental results show that the *topic cluster* constraint improves clustering cohesion and stability.

In this task, we first use multilingual-sentence-bert [16] to encode text. It achieves promising performance in the task of topic detection.

The contributions of this work are summarized below:

- We propose the concept of *topic cluster* and a stream clustering algorithm to improve the accuracy of fine-grained topic clustering. In addition, our out time weighting mechanism for news can effectively distinguish news that occur in different times.
- Experimental results show our method achieving remarkable performance.

2   Related Work

Researchers have investigated a variety of methods for stream clustering [1,11]. Many algorithms like CluStream [2], StreamKM++ [1] are partition-based algorithms. In these algorithms, the number of clusters have to be predefined. This is not suitable for news topic detection. Though the algorithm is simple, it only restricts the distance between each sample and the cluster center, resulting in dissimilar samples being gathered together.

Grid based algorithms use the grid data structure, which divides the whole space into a number of cells. Then, these cells are clustered to form the clustering result.

Most density-based clustering methods adopt the online-offline paradigm. The paradigm tracks up-to-date news in real time online and calculates the clustering result offline. Based on this paradigm, researchers proposed many methods including, DenStream [6] and D-Stream [7].

Another widely used type of clustering methods is hierarchical clustering, which generates clusters by iteratively combining the closest or most similar two clusters. It has a very popular successor, BIRCH [18], which performs better in terms of time efficiency. Another type of aggregative clustering is the SinglePass algorithm [5].

Providing an appropriate text representation for topic detection is a challenging problem. With the development of deep learning, Word2Vec [13] was proposed to learn high-quality distributed vector representations of word embedding. Sentence-Bert [15] uses siamese and triplet network structures to derive semantically meaningful sentence embeddings that can be compared using cosine-similarity. In our experiment, we compared different text representations in detail and the multilingual-sentence-bert representation performed the best.

3 Preliminaries

3.1 Task Formulation

We shall first explain some theoretical notions by defining the concepts. We formulate the problem of detecting fine-grained topics after introducing these concepts to readers: given a news stream $G = \{s_1, s_2, s_3...s_i\}$, the goal is to aggregate news into different topics $C = \{c_1, c_2, c_3...c_k\}$ incrementally in real time. According to the occurrence of events, we put these outdated topics in the topic cluster queue C to $O = \{c_1, c_2, c_3...c_n\}$ to make C more efficient.

3.2 Distance Definition

We introduce the definitions of distance between (1) sample and cluster (2) two different samples. We adopt multilingual-sentence-bert [16] as the encoder to generate the embeddings of news data. The model takes a piece of text sequence as input and outputs a fixed dimension vector. The embedding of the cluster center is defined as the average embedding value of all the samples in the cluster.

$$c_k = \frac{\sum_{n=1}^{N} s_n}{n} \qquad (1)$$

s_n is the sample in c_k. Then the similarity between s_i and c_k

$$cs_{ki} = \frac{c_k \cdot s_i}{\|c_k\| * \|s_i\|}, \qquad (2)$$

s_i is the document embedding arriving at time i, c_k is the kth cluster in the topic cluster queue. Similarly, the similarity between two samples is

$$ss_{ik} = \frac{s_i \cdot s_k}{\|s_i\| * \|s_k\|} \qquad (3)$$

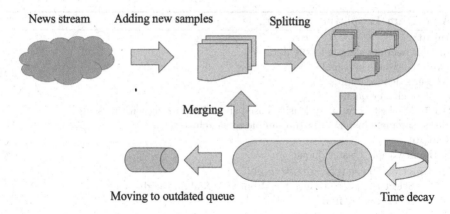

News stream   Adding new samples   Splitting

Merging

Moving to outdated queue   Time decay

Fig. 1. Model framework

In addition, we use the time decay function to characterize the process in which similarity of the articles change with time difference.

$$\gamma = e^{(\frac{-1*(|t_i - t_j|)}{h})^p * \log 2},$$
(4)

where γ is the similairty of two times, t_i is the time document s_i arrives, t_j is the time when cluster c_j was created, h and p are two parameters that practically set by 15 and 1.8. In the experiment, we get the final distance between news s_i and cluster c_k

$$s_{ki} = 1 - \gamma * cs_{ki},$$
(5)

s_{ki} is the distance to identify where we should insert the new sample.

4   Algorithm

In this section, we will go into detail and introduce our proposed Iterative Strict Density-based Clustering (ISDC). Figure 1 is our model framework. Our clustering algorithm consists of two parts: (1) Dynamic topic detection management (2) Outdated topic detection management.

4.1   Dynamic Topic Detection Management

Adding New Samples. To discover clusters in an evolving news stream, we maintain a dynamic queue of *topic clusters*. When a new sample s arrives, the procedure of inserting it into the topic cluster queue is described below: 1. First we calculate the distances between the new sample and all existing clusters. If the distance to the nearest cluster is below our threshold δ, we insert the new sample into the nearest topic cluster c_p. Else we create a new topic cluster for this new sample.

Algorithm 1. Splitting and merging

Input:

s_i: next pending sample in the data flow

δ:user's defined threshold

ϵ_0: user's defined radius

C: topic cluster queue

DBSCAN(cluster, radius): density-based spatial clustering with noise[8]

Merge(sample, cluster): add the sample to the cluster.

Output:

S: the new topic cluster queue

```
 1: function "Iterative clustering"
 2:     Select the nearest cluster c_k from C, calculate the distance d_i
 3:     if  d_i <= δ  then
 4:         Add s_i to the cluster c_k
 5:         if c_k is not a topic cluster then
 6:             ε_0 ← ε_0 - 0.01
 7:             sub ← DBSCAN(cluster=c_k,radius = ε_0)
 8:             iso ← c_0 \ sub
 9:             for s ∈ sub do
10:                 if s is not a topic cluster then
11:                     goto Line 5
12:                 for p ∈ iso do Merge(sample=p,cluster=sub)
13:             add sub to S
14:     return S
```

Splitting and Merging. We set the cosine distance threshold as δ. As news event s_i arrives, we compare the distance between x_i and all existing topic clusters. After calculating those distances, we find the shortest cosine distance d_i between s_i and all topic clusters. If d_i is less than δ, we add x_k to the nearest cluster, else we initialize a new topic cluster with x_k. We restrict all clusters to be topic clusters, but the modified cluster is likely to violate the criteria due to inclusion of the new document. Hence, we check whether the modified cluster is still a topic cluster. If not, we shall split the cluster into several *topic clusters*.

In our method, we use DBSCAN [8] to split the modified cluster. DBSCAN defines clusters as the largest collection of densely connected points, it can divide regions by distance threshold. Let origin cosine distance threshold be the predefined ϵ_0. If the cluster isn't a topic cluster, in practice, we decrease ϵ_0 to ϵ_1 by 0.01. The aim is to divide the original cluster into more cohesive sub-clusters. If sub-clusters satisfy the criterion to be topic clusters, we stop splitting. If subclusters still fail to be topic clusters, then we have to iteratively split the subclusters until they become topic clusters. The number of iterations is constant because the lower limit of the cosine distance is 0. Since the number of iterations is constant, the time complexity of such iterations is $O(n)$, an acceptable bound.

After splitting, the origin cluster is divided into several topic clusters. However, there are some isolated samples that may belong to the topic clusters. Hence we need to detect isolated samples and check if they belong to a topic cluster. We still use our previous methods of splitting: First calculate the distances between

Algorithm 2. Time Decay

Input:
U: the outdated queue
decay:decay function
C: the topic cluster queue
p: user's defined time period
w: user's defined weight threshold

1: **function** TIME DECAY
2:     **for** every time interval of p **do**
3:         **for** c_k in C **do** time weight of c_k, $w_k = decay(w_k)$
4:             **if** $w_k < w$ **then**
5:                 move c_k to U

the isolated samples and topic clusters. If the distance is less than the original threshold δ_0, we add the sample to the cluster, else we initialize a new cluster for the isolated sample.

4.2 Outdated Topic Detection Management

Time Decay. For each existing topic cluster c_p, the weight will decay over time. The decay function is

$$w_p = w_p * 2^{-1*\sigma} \tag{6}$$

σ is the attenuation coefficient.

If w_p is less than w, it means that the cluster is outdated and should be moved to the outdated cluster queue. The outdated cluster queue is used to store all the history news events outside of the current time window. We periodically check and reduce the weight of the topic cluster. An important problem is how to determine the value of this time period. Generally speaking, news reports on a specific news event rarely last more than five days. Thus we set the period value as four or five days.

Moving to Outdated Queue. Unlike our algorithm, many other stream clustering algorithms adopt a different strategy. They adopt the online-offline strategy which store snapshots of the data stream and computes clustering results when necessary. Our method, on the contrary, combines the two steps into one. We directly compute the final clusters and adjust them dynamically. Then we merge the topic cluster queue and the outdated queue to get the final clustering result. This way, we reduce the computational pressure of the offline clustering process, and thus the calculation of the whole process is more balanced.

5   Experiment

5.1 Performance Metrics

A good clustering algorithm requires all clusters to have high intra-cluster similarity and low inter-cluster similarity. Here are some common metrics in

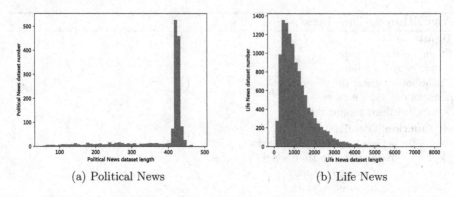

(a) Political News   (b) Life News

Fig. 2. News length comparison.

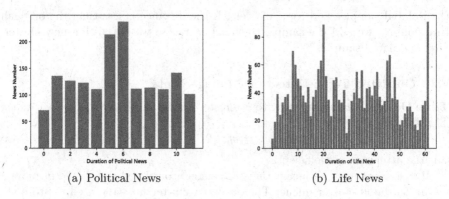

(a) Political News   (b) Life News

Fig. 3. News time duration comparison.

clustering. We use Purity, Silhouette Coefficient [4], FMI (Fowlkes–Mallows index) [9], and V-M. (V-measure) [17] as evaluation metrics for clustering results. Purity calculates the proportion of correctly clustered documents in total documents. FMI describes the difference between clustering result and the ground truth. V-measure is the harmonic mean of homogeneity and completeness, it comprehensively reflects the overall performance of the algorithm.

5.2   Dataset Analysis

To ensure the comprehensiveness of our experiment, we experimented on two test datasets with completely different data distributions. We made a detailed comparsion of the two datasets in Fig. 2 and Fig. 3 The first news dataset is from a data mining system NewsMiner [10], an online news discovery and mining website. We selected all documents about political figures in the U.S. in 7 days, then our experts manually classified them into different categories.

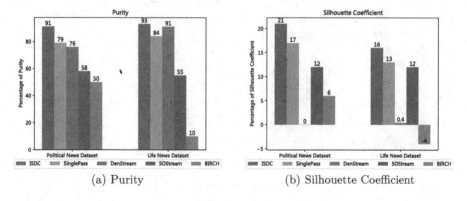

(a) Purity   (b) Silhouette Coefficient

Fig. 4. Cluster cohesion comparison

The second test set is from *Growing story forest online from massive breaking news* [12]. The total number of news is 11748, with an average length of 1210.5 words. This news set contains news in many fields, including finance, sports, weather forecast, etc.

The length of each news piece in the Political News Dataset is concentrated in around 400–450 words, while lengths of news pieces in the Life news Dataset are more scattered, with lengths ranging from 0 to 5000 words.

5.3   Comparison with Baseline Algorithms

Parameter Settings. We experimented on the two datasets to compare different clustering methods. We compared our method with 4 clustering algorithms including BIRCH [18], SinglePass [14], DenStream [6], SOStream [11]. The common parameters in the experiment are a) T, the cluster merge threshold used in density-based algorithms, b) N the predefined cluster number used in hierarchical clustering algorithms. We optimize these parameters separately using grid search. Grid search not only ensures that comparisons between the unsupervised methods are fair, but also gets the best achieveable results of each individual method. Optimal results for each method are shown in the chart below.

Experimental Results. Our experimental strategy consists of two steps: First we run different clustering methods and obtain their respective aggregated samples. Then, we assign a new cluster to every isolated sample. For fairness of comparison, we use multilingual-sentence-bert [16] as the embedding model for all clustering algorithms.

Figure 4 displays a comparison of the clustering algorithms in terms of purity and Silhouette Coefficients.

Purity is defined as

$$Purity(\Omega, C) = \frac{1}{N} \sum_k \max_j |\omega_k \cap c_j| \tag{7}$$

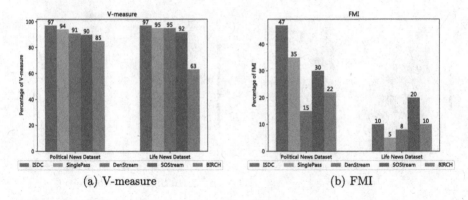

(a) V-measure   (b) FMI

Fig. 5. Comprehensive performance comparison.

N is the total number of samples, $\Omega = \{w_1, w_2, ..w_K\}$ is the predicted cluster set. $C = \{c_1, c_2, ..c_J\}$ is the true cluster set. We can see from the formula: the higher the purity, the better cohesion within the predicted cluster. Our ISDC strictly controls the distance between samples, so our method outperformed its counterparts in terms of purity on these two datasets, as expected.

In order to exclude the impact of the annotated labels, we adopted an unsupervised indicator: the Silhouette Coefficient

$$SC = \frac{1}{N} \sum_{i=1}^{N} SC(d_i) \tag{8}$$

$$SC(d_i) = \frac{b - a}{max(a, b)} \tag{9}$$

SC is the total Silhouette Coefficient, $SC(d_i)$ is the Silhouette Coefficient of cluster i. a is the average distance between a sample and other samples in its cluster, and b is the average distance between a sample and other cluster samples. The larger the Silhouette Coefficient is, the more compact the instances in the cluster are. Our clustering algorithm produces the most accurate reflection of difference in text embedding.

Figure 5 compares the algorithms with regard to two comprehensive indicators: V-Measure and FMI score. V-Measure is completely based on the conditional entropy between the two clusters, that is, after a certain category is divided, the uncertainty of the other category is determined. The smaller the uncertainty, the closer the two categories are divided. Therefore, the corresponding h value or c value is greater. V-measure is the harmonic mean of homogeneity and completeness, and it can more comprehensively reflect the effect of clustering. Due to our maintenance of the topic cluster queue, our method slightly outperforms others in terms of V-measure.

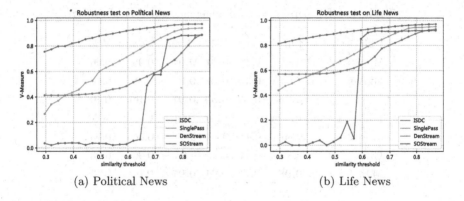

(a) Political News   (b) Life News

Fig. 6. Cluster robust comparison

The Fowlkes-Mallows Index (FMI) [9] is defined as the geometric mean of the pairwise precision and recall:

$$FMI = \frac{TP}{\sqrt{(TP+FP)(TP+FN)}} \tag{10}$$

where TP is the number of true positives, FP is the number of false positives, and FN is the number of false negatives. Here the SOStream algorithms performs best among all algorithms. This is because FMI encourages samples to generate large clusters, and SOStream, with its loose clustering standards, performs well on this indicator.

In order to test the robustness of ISDC, we tested the performance of four algorithms under gradually changing parameters in Fig. 6. Different algorithms have different robustness. The SOStream's performance depends on the correctness of the parameter, and small similarity threshold will lead to a sharp decline in the performance of the algorithm. The SinglePass and DenStream algorithms have poor performance when the similarity threshold is small, but they have a rising performance when the similarity threshold is bigger. On contrast, our algorithm can stably achieve the optimal effect in different parameters, which can be seen as strong robustness.

5.4   Ablation Study on the Embedding Model

We tried three embedding models for our news topic detection task in Table 1. While using ISDC as our clustering algorithm, we tested Word2Vec, GloVe, M.S.(multilingual-sentence-bert) and compared their performances. The Word2Vec model and the GloVe model were pretrained on the news corpus in our Newsminer system.

Based on BERT, the multilingual-sentence-bert (M.S.) was fine-tuned with STS (Semantic Textual Similarity) and NLI (natural language inference). Table 1

Table 1. ISDC performace with different sentence embedding

Chinese news	Purity	AMI	FMI	Homo.	Comp.	V-M.
Word2Vec	0.83	0.32	0.20	**0.97**	0.95	0.96
GloVe	0.83	0.32	0.19	**0.97**	0.95	0.96
M.S.	**0.96**	**0.42**	**0.35**	**0.97**	**0.99**	**0.98**
Political news	Purity	AMI	FMI	Homo.	Comp.	V-M.
Word2Vec	0.86	0.43	0.28	0.95	0.96	0.95
GloVe	0.87	0.44	0.27	0.95	0.96	0.95
M.S.	**0.98**	**0.76**	**0.61**	**0.97**	**0.99**	**0.98**

shows that Word2Vec and GloVe have almost the same performance, while Sentence Transformer clearly outperforms the two. The M.S. produces even more exceptional results under the metrics of Purity, AMI and FMI.

6   Conclusion

In this paper, we describe the task of News Topic Detection (NTD). To accomplish this task, we describe a clustering algorithm which can generate *core clusters* by iteratively using the DBSCAN algorithm. In comparison to other baseline algorithms, our method achieved outstanding performance on the NED task while maintaining a simple structure. We also made a detailed discussion about the performance of ISDC and other algorithms. Our method is more robust and performs better. In the ablation study on the Embedding model, we found that the multilingual-sentence-bert [16] has a significant advantage over Word2Vec and GloVe. Although we have achieved promising experimental results, accuracy problems do occur when text representation of news is not accurate enough. We will further explore this issue in our future works.

Acknowledgements. This work is supported by the NSFC Key Project (U1736204) and grants from the Institute for Guo Qiang, Tsinghua University (2019GQB0003) and HUAWEI Inc.

References

1. Ackermann, M.R., Märtens, M., Raupach, C., Swierkot, K., Lammersen, C., Sohler, C.: StreamKM++ a clustering algorithm for data streams. J. Exp. Algorithmics (JEA) **17**, 2-1 (2012)
2. Aggarwal, C.C., Philip, S.Y., Han, J., Wang, J.: A framework for clustering evolving data streams. In: Proceedings 2003 VLDB Conference, pp. 81–92. Elsevier (2003)
3. Allan, J., Carbonell, J.G., Doddington, G., Yamron, J., Yang, Y.: Topic detection and tracking pilot study final report (1998)

4. Aranganayagi, S., Thangavel, K.: Clustering categorical data using silhouette coefficient as a relocating measure. In: International Conference on Computational Intelligence and Multimedia Applications (ICCIMA 2007), vol. 2, pp. 13–17. IEEE (2007)

5. Baeza-Yates, R., Ribeiro-Neto, B., et al.: Modern Information Retrieval, vol. 463. ACM Press, New York (1999)

6. Cao, F., Estert, M., Qian, W., Zhou, A.: Density-based clustering over an evolving data stream with noise. In: Proceedings of the 2006 SIAM International Conference on Data Mining, pp. 328–339. SIAM (2006)

7. Chen, Y., Tu, L.: Density-based clustering for real-time stream data. In: Proceedings of the 13th ACM SIGKDD International Conference on Knowledge Discovery and Data Mining, pp. 133–142 (2007)

8. Ester, M., Kriegel, H.P., Sander, J., Xu, X.: Density-based spatial clustering of applications with noise. In: International Conference on Knowledge Discovery and Data Mining, vol. 240, p. 6 (1996)

9. Fowlkes, E.B., Mallows, C.L.: A method for comparing two hierarchical clusterings. J. Am. Stat. Assoc. **78**(383), 553–569 (1983)

10. Hou, L., Li, J., Wang, Z., Tang, J., Zhang, P., Yang, R., Zheng, Q.: NewsMiner: multifaceted news analysis for event search. Knowl.-Based Syst. **76**, 17–29 (2015)

11. Isaksson, C., Dunham, M.H., Hahsler, M.: SOStream: self organizing density-based clustering over data stream. In: Perner, P. (ed.) MLDM 2012. LNCS (LNAI), vol. 7376, pp. 264–278. Springer, Heidelberg (2012). https://doi.org/10.1007/978-3-642-31537-4_21

12. Liu, B., Niu, D., Lai, K., Kong, L., Xu, Y.: Growing story forest online from massive breaking news. In: Proceedings of the 2017 ACM on Conference on Information and Knowledge Management, pp. 777–785 (2017)

13. Mikolov, T., Sutskever, I., Chen, K., Corrado, G.S., Dean, J.: Distributed representations of words and phrases and their compositionality. In: Advances in Neural Information Processing Systems, pp. 3111–3119 (2013)

14. Papka, R., Allan, J., et al.: On-line new event detection using single pass clustering. Univ. Massachusetts Amherst **10**(290941.290954), 1–10 (1998)

15. Reimers, N., Gurevych, I.: Sentence-BERT: sentence embeddings using siamese bert-networks. arXiv preprint arXiv:1908.10084 (2019)

16. Reimers, N., Gurevych, I.: Making monolingual sentence embeddings multilingual using knowledge distillation. In: Proceedings of the 2020 Conference on Empirical Methods in Natural Language Processing. Association for Computational Linguistics, November 2020. https://arxiv.org/abs/2004.09813

17. Rosenberg, A., Hirschberg, J.: V-measure: a conditional entropy-based external cluster evaluation measure. In: Proceedings of the 2007 Joint Conference on Empirical Methods in Natural Language Processing and Computational Natural Language Learning (EMNLP-CoNLL), pp. 410–420 (2007)

18. Zhang, T., Ramakrishnan, R., Livny, M.: BIRCH: an efficient data clustering method for very large databases. ACM SIGMOD Rec. **25**(2), 103–114 (1996)

19. Zubaroğlu, A., Atalay, V.: Data stream clustering: a review. arXiv preprint arXiv:2007.10781 (2020)

A Pre-LN Transformer Network Model with Lexical Features for Fine-Grained Sentiment Classification

Kaixin Wang, Xiujuan Xu[✉], Yu Liu, and Zhehuan Zhao

School of Software, Dalian University of Technology, Dalian 116620, Liaoning, China
xjxu@dlut.edu.cn

Abstract. Sentiment classification is an important task of sentiment analysis, which aims to identify different sentiment polarity in subjective text. Although most existing models can effectively identify the extreme polarity (extremely positive, extremely negative), we find they cannot distinguish the intermediate polarity (generally positive, neutral, generally negative) clearly. Besides, the models based on convolutional neural networks (CNNs) and recurrent neural networks (RNNs) also have some problems, such as weak parallel computing power and poor long-distance dependence capacity. This paper proposes a new model based on Pre-LN Transformer and lexical features, which can improve fine-grained sentiment classification of online reviews. In this work, the Pre-LN Transformer encoder with multi-headed self-attention captures hidden features of different subspaces. Unlike the Post-LN Transformer, the Pre-LN Transformer places the normalization layer in the residual block to make the model more stable. On this basis, we reconstruct the Vader lexicon and further integrate sentiment lexical features extracted from the lexicon into the model. We perform sentiment classification tasks on two publicly available online review datasets. Experimental results show that our model achieves state-of-art performance while distinguishing fine-grained sentiment.

Keywords: Sentiment classification · Pre-LN transformer · Lexical features · Fine-grained

1 Introduction

Online reviews play an important role in e-commerce. Previous research shows that when customers make a transaction, they tend to evaluate a product or service by browsing online reviews, and then decide whether to buy the product or not [1]. Analyzing and identifying the emotions expressed by consumers in reviews can provide personalized services for consumers and help enterprises to draw up marketing strategies. Sentiment classification is a subtask of sentiment analysis, which can automatically identify the emotional tendency in subjective text.

Some previous methods have been proposed for sentiment classification. However, there are some shortcomings in the existing models. First, most of these models use language models with convolutional neural networks [2–4] or recurrent neural networks [5,

© Springer Nature Switzerland AG 2021
H. Lin et al. (Eds.): CCIR 2021, LNCS 13026, pp. 100–111, 2021.
https://doi.org/10.1007/978-3-030-88189-4_8

6]. The recurrent structure has obvious parallelism defects, which leads to low computational efficiency. Convolutional neural network is not good at dealing with long-distance dependence, while online reviews are mainly focused on sentence-level and document-level text, so it has higher requirements for remote dependence. In addition, attention mechanism began to be widely used in sentiment analysis, which can solve the problem of long-distance dependence and improve the parallel computing ability of the model. In particular, the Transformer based model uses multi-headed attention mechanism, which allows the model to learn information from different subspaces. But the training and hyperparameter adjustment of the existing Transformer based models [7, 8] take a lot of time. What's more, the neural networks alone cannot distinguish the subtle emotional differences well. Most existing sentiment classification models can effectively identify the extreme polarity (extremely positive, extremely negative), but cannot distinguish the intermediate polarity (generally positive, neutral, generally negative) clearly.

To overcome the limits of the previous methods, we propose a model for fine-grained sentiment classification based on Pre-LN Transformer network and lexical features. In detail, word2vec is used to train the word vectors to obtain the superficial semantic features, then we capture hidden features through the encoder in Pre-LN Transformer [9]. Furthermore, we reconstruct VADER [10] lexicon to form the emotion lexicon in the field of food. Next, we use it to quantify the sentiment intensity of the text and distinguish the subtle differences of emotion, and further extract the sentiment vector of reviews. Finally, we integrate the sentiment vector into the deep network structure features to achieve the fine-grained sentiment classification.

The main contributions of this work are summarized as follows:

- We propose a model based on Pre-LN Transformer network, and integrate sentiment lexical features into the deep network to realize fine-grained sentiment classification.
- We reconstruct VADER lexicon and divide VADER emotion into five detailed levels to quantify the emotional intensity of the text, which can improve the ability to recognize the subtle differences of emotion.
- We perform comprehensive experiments on publicly available online review datasets and our model achieves state-of-art performance compared with other methods.

2   Related Work

Sentiment classification mainly includes rule-based methods [11], machine learning methods [12, 13] and deep learning methods [14]. Convolutional neural networks (CNNs), recurrent neural networks (RNNs) and their variants have been widely used in sentiment classification tasks. Specifically, Santos et al. [3] used FastText word embedding as word representation combined with convolutional neural network to complete the task of sentiment analysis. Hameed et al. [15] proposed a deep learning model for binary sentiment classification based on a single-layer Bi-directional Long Short-Term Memory network, which adopted an optimization strategy for the pooling layer. However, RNNs and CNNs have disadvantages of weak parallel computing power and poor long-distance dependence ability respectively.

With the proposal of attention mechanism [16], the combination of neural network and attention mechanism has become the mainstream method. Yang et al. [17] effectively used hierarchical attention network to solve the problem of long sentence classification. Zheng et al. [5] proposed a hybrid bidirectional recurrent convolutional neural network, which used attention mechanism to capture key components of text. However, the traditional attentional mechanism leads to the dependence on external information.

Previous studies have promoted the development of sentiment classification, but there are also some challenges. Transformer is applied to sentiment analysis as a non-pre-trained model to solve the above problems. Experimental verification shows that Transformer is superior to recurrent neural network or convolutional neural network in terms of long-distance feature capture and comprehensive feature extraction [18], and it also has obvious advantages in parallel computing [19]. The Transformer structure uses multi-headed attention mechanism to allow models to learn relevant information in different presentation subspaces. The output of Transformer does not rely on the computation of the previous moment, which greatly improves the parallel computing ability of the model. However, the existing Transformer model using post-layer normalization (Post-LN) [20] is sensitive to parameter adjustment, and it takes a long time to train the model. Recently, the PreLN Transformer structure has been proposed by Xiong et al. [9] to solve these problems. They put the normalization layer in the process of residual block to make the training more stable, which inspired us to use the Pre-LN Transformer for sentiment classification. What's more, we find that only using independent neural networks cannot distinguish fine-grained sentiment well. We believe that learning emotion from text has an important relationship with lexical features, just as humans begin to learn language from dictionaries. The above research and challenges inspire our work.

3   Model

The model structure is shown in Fig. 1. Next, we introduce all the components of the model from bottom to top.

The problem can be defined as follows. There is a label set Y containing 5 category labels $Y = \{Neg-, Neg, Neu, Pos, Pos+\}$, which represent extremely negative, generally negative, neutral, generally positive, and extremely positive emotional intensity respectively. Given an input sequence S containing n words, denotated as $S = \{w_1, w_2, ...w_n\}$. Sentiment classification task aims to predict the emotion intensity of sequence S in label set Y.

3.1   Input Layer

The input layer is the beginning of the model and the input text is converted into index sequences.

3.2   Embedding Layer

Word Embedding. Based on our corpus, word2vec is used to train the word vector to form an embedded matrix $\mathbb{R}^{v \times d_1}$, where v is the vocabulary size of dataset and d_1 portrays the dimension of word embedding. The input word index is mapped to the embedding matrix, transforming into a dense vector. Computational Linguistics We encode each

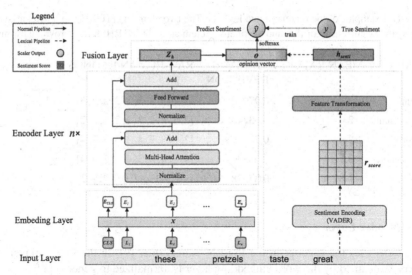

Fig. 1. Model architecture. The word embedding of the review is shown in light blue. The sentiment lexical features of the review are shown in purple. The model uses the Encoder in the Pre-LN Transformer to get information for each word, which is shown in orange. (Color figure online)

input position to obtain the position information when calculating the dot product of attention, represented by L_n in Fig. 1. The embedding vector of the n-th word w_n in text S can be expressed as $x_n \in \mathbb{R}^{d_1}$. Embedding layer is represented as shown in formula (1).

$$X = [x_1, x_2, \dots x_n] \tag{1}$$

Sentiment Embedding. VADER [10] sentiment analyzer is used to encode sentiment embedding. Based on the original VADER lexicon, we divide the reviews into words and count the top 2,000 words with the highest frequency in the reviews. We first find 10 independent people to evaluate each word which we want to add to the lexicon and make sure the standard deviation is no more than 2.5. Then we select 271 emotional words that can express subjectivity from the field of food and mark them from -4 (extremely negative) to 4 (extremely positive). Some of the emotion words added are shown in Table 1.

We change the original three typical thresholds of "positive", "neutral" and "negative" emotion of VADER into five levels, including "extremely positive" ($0.6 \leq$ compound ≤ 1), "relatively positive" ($0.2 \leq$ compound < 0.6), "neutral" ($-0.2 \leq$ compound < 0.2), "relatively negative" ($-0.6 \leq$ compound < -0.2), and "extremely negative" ($-1 \leq$ compound < -0.6). The compound score is calculated by the sum of the rule-adjusted valence scores of each word in the lexicon, and then normalize them to -1 between (extremely negative) and 1 (extremely positive). For a review, r_{score} represents the sentiment score given by VADER. h_{senti} represents the sentiment vector obtained from the transformation of r_{score}, covering the sentiment lexical features of the review text.

Table 1. Examples of the reconstructed lexicon. The two elements (TOKEN and MR) are used directly by the current algorithm and the final two elements (SD and RHSR) are provided for rigor.

TOKEN	MEAN-RATING	STANDARD DEVIATION	RAW-HUMAN-RATINGS
Cheap	2.2	0.97980	[3, 0, 3, 3, 1, 3, 2, 2, 2, 3]
Convenient	2.0	0.63246	[2, 1, 3, 2, 3, 2, 2, 1, 2, 2]
Freshly	2.9	0.7	[2, 3, 3, 2, 4, 3, 3, 3, 2, 4]
Impeccable	3.6	0.66332	[4, 2, 4, 4, 3, 4, 4, 3, 4, 4]
Tasteless	−1.8	0.87178	[−2, −1, −2, 0, −1, −2, −3, −2, −2, −3]

3.3 Encoder Layer

Our model uses the encoder in the Pre-LN Transformer [9]. In the first sublayer, the matrix X obtained by the word embedding layer is normalized to generate X^N.

$$X^N = LayerNorm(X) \tag{2}$$

The matrix X^N is multiplied with different weight matrices to obtain Q, K and V, which represent query vector, key vector and value vector respectively. The calculation is as follows:

$$Q = X^N W^Q \tag{3}$$

$$K = X^N W^K \tag{4}$$

$$V = X^N W^V \tag{5}$$

where W^Q, W^K, $W^V \in \mathbb{R}^{d_1 \times d_2}$ are different weight matrices.

Multi-Head Attention improves the ability of the model to pay attention to different positions and gives the attention layer several representation subspaces. Equations (6) and (7) give the general form of the function to calculate this value.

$$H = head_1 \oplus head_2 \oplus \ldots \oplus head_m \tag{6}$$

$$head_i = softmax\left(\frac{Q_i K_i}{\sqrt{d_k}}\right) V_i \tag{7}$$

where m is the number of heads, $H \in \mathbb{R}^{n \times d_2 m}$ and $\oplus$ denotes concatenation operation.

Where Q_i, K_i, V_i represent the query direction, key vector and value vector of different headers, $I \in [1, m]$. $\sqrt{d_k}$ is the regulatory factor which plays a regulating role to make the inner product result not too large. The output of multi-head attention layer is shown in formula (8).

$$Z_{att} = MutiHead(Q, K, V) = HW^O \tag{8}$$

where $W^O \in \mathbb{R}^{d_2 m \times d_1}$ is the weight matrix of the multi-head attention. Transformer uses residual connections to avoid gradient loss in deep networks, which can be simplified into the following form:

$$Z = X + Z_{att} \tag{9}$$

In the second sublayer of the Encoder, the Z is normalized to get Z^N, which is then input into the feedforward neural network to get the hidden vector Z_r. Feedforward neural network is a two-layer fully connected layer. The activation function of the first layer is Relu, and the second layer does not use activation function. Finally, Z_h was obtained by residual connection. The formula is as follows:

$$Z^N = LayerNorm(Z) \tag{10}$$

$$Z_r = \text{Relu}(Z^N W^1 + b_1) W^2 + b_2 \tag{11}$$

$$Z_h = Z + Z_r \tag{12}$$

where W^1, W^2 are the weight matrix of feedforward network and b_1, b_2 are the bias.

3.4 Fusion Layer and Objective Function

We connect the output of the encoder layer after dimensionality reduction with the sentiment vector h_{senti}, and obtain the opinion vector as shown in Eq. (13). Softmax is used to calculate the sentiment targets distribution of reviews:

$$o = [Z_h, h_{\text{senti}}] \tag{13}$$

$$P(y_i) = \text{softmax}\left(W_{\text{tag}} \cdot o + b_{\text{tag}}\right) \tag{14}$$

where W_{tag} maps the opinion vector o to the feature score of each emotional tag, and b_{tag} is the bias item. In an iteration, the loss function is defined as follows:

$$L(\theta) = \frac{1}{N}\left[\sum_{i=0}^{N} -y_i \log p(y_i) + \lambda R(\theta)\right] \tag{15}$$

where $p(y_i)$ is the prediction, y_i is the true target, N is the number of samples and λ denotes the coefficient for L2 regularizer $R(\theta)$.

4 Experiments

4.1 Datasets

We conduct our experiment on two publicly available datasets: Yelp review datasets[1] and Amazon food reviews. The Yelp dataset consists of online reviews from restaurant customers, from which we extract the reviews of restaurants in Las Vegas. In the Amazon dataset, we select the 2012 food category reviews of Amazon website [21].

After data preprocessing, Table 2 shows the statistics and category for each dataset. And then the two datasets are divided into training set, verification set and test set according to the ratio of 6:2:2, as shown in Table 3.

---

[1] https://www.yelp.com/dataset_challenge.

Table 2. Statistics of used datasets. Neg−, Neg, Neu, Pos and Pos+ respectively represent five degrees of emotion from extremely negative to extremely positive.

Dataset	Neg−	Neg	Neu	Pos	Pos+	Total
Yelp	19745	17141	27197	49738	67496	181317
Amazon	20368	11208	15603	28938	122542	198659

Table 3. Partitioning of single modal datasets

Dataset	Train	Valid	Test	Total
Yelp	108789	36264	36264	181317
Amazon	119195	39732	39732	198659

4.2 Evaluation

To evaluate the effect of our model, we use $MicroF_1$ and $MacroF_1$ as the evaluation standard. Considering the evaluation method of classification effect in the case of multi-class, we evaluate Micro averaging and Macro-averaging respectively. Among them, $MicroF_1$ is more susceptible to the largest number of classes, $MacroF_1$ can treat each class equally.

4.3 Training Details

In the experiments, we set the max length is 200. We use word2vec to train the word vector to form an embedding matrix, in which the dimension of word embedding is set to 128. We set the hyperparameters to train our model. For the encoder in the Pre-LN Transformer, we set the number of heads in the multi-head self-attention mechanism to 8. In the full connection layer, we employ dropout value is 0.5 to prevent overfitting. The batch size and learning rate are set to 32 and 0.001 respectively. During the training process, Adam optimizer is used to optimize the model.

4.4 Comparison Methods

To validate the performance of our model, we compare it with several benchmark approaches, including some shallow neural networks, convolutional neural networks and their variants (CNNs), recurrent neural networks and their variants (RNNs), and Transformer based network models. The comparison method used in the experiment are described as follows:

TextCNN [22] is a method for text classification using convolutional neural network.

FastText [23] is a fast text classifier that uses average word or n-gram em-bedding for document embedding.

HAN [17] utilizes hierarchical attention mechanism based on bidirectional recurrent neural network for text classification.

BiLSTM [24] consists of two separate LSTMs, which can read the input word sequence forward and backward to get more context information.

AM-Bi-LSTM [25] uses BILSTM to learn text representation, and then applies the attention mechanism to dynamically assign weight to words.

Single-layered BiLSTM [15] is a deep learning model for sentiment classification based on BiLSTM, which adopts optimization strategy with a global pooling mechanism.

Post-LN Transformer [20] leverages the encoder in Transformer to capture the hidden features of text to improve parallel computing power, and the multi-head attention mechanism in the encoder can learn the features of different subspaces.

Pre-LN Transformer [9] puts the normalized layer in Transformer before the multi-head attention mechanism layer and the feedforward neural network respectively, which is different from the Post-LN Transformer.

SPOLNT: In this model, Post-LN Transformer is used to extract text features and Vader sentiment analyzer is used to extract sentiment lexical features. Then, the text features and sentiment lexical features are fused and input into classifier for sentiment classification.

SPRLNT: The main methods proposed in this paper.

4.5 Results

Table 4. Results of all proposed models on Yelp and Amazon datasets in terms of $MicroF_1$ and $MacroF_1$ score (%)

模型	Amazon		Yelp	
	$MicroF_1$	$MacroF_1$	$MicroF_1$	$MacroF_1$
FastText	65.91	34.44	52.70	43.56
TextCNN	68.11	38.20	53.58	42.66
BiLSTM	71.77	46.02	59.05	51.62
HAN	73.83	48.89	60.58	54.20
AM-Bi-LSTM	73.18	51.84	60.74	55.85
Single-layered BiLSTM	73.42	53.95	60.82	56.65
Post-LN transformer	74.58	52.38	61.15	54.80
Pre-LN transformer	74.56	55.23	61.55	56.61
SPOLNT	**74.69**	53.31	61.17	55.90
SPRLNT	74.29	**58.66**	**61.89**	**57.73**

We list the results of different models for sentiment classification on Amazon and Yelp datasets, as shown in Table 4. From the experimental results, our model achieves the best performance in the evaluation of $MicroF_1$ and $MacroF_1$ on Yelp dataset. Among the baseline models (TextCNN, FastText, HAN and BiLSTM), HAN performs better, which indicates that GRU can better capture the Bi-directional semantic dependence of text and hierarchical attention mechanism can focus on the important emotional information.

As we can see, Transformer based models perform better than most convolutional and recurrent neural network models because Transformer uses multiple attention mechanisms to learn relevant emotional information in different presentation subspaces. Compared with the Pre-LN Transformer, the $MacroF_1$ score of SPRLNT on Amazon and Yelp datasets are increased by 3.43% and 1.12%, which indicates that the Vader lexicon can distinguish the nuances of emotions by quantifying the emotional intensity of the text. It can be found that the $MicroF_1$ of SPRLNT on Amazon dataset is slightly lower than that of SPOLNT, but it is higher than that of SPOLNT on other three evaluation. In particular, SPRLNT improves the performance by 5.35% and 1.83% in $MacroF_1$ evaluation than SPOLNT on Amazon and Yelp datasets, respectively. It shows that Pre-LN Transformer changes the position of normalization layer, which not only makes the training more stable, but also can effectively extract the deep semantic features of the text.

Table 5. Results based on the fusion of sentiment features with different baseline models

Model	$MicroF_1$	Δ	$MacroF_1$	Δ
Sen-FastText	54.15	↑1.45	44.31	↑0.75
Sen-TextCNN	54.20	↑0.62	45.93	↑3.27
Sen-BiLSTM	58.98	↓0.07	53.53	↑1.91
Sen-HAN	61.02	↑0.44	55.77	↑1.57

Table 5 lists the sentiment classification results on Yelp dataset after different baseline models combine sentiment lexical features, where "Δ" represents the promotion effects of sen-model respectively relative to the four baseline models without emotional features (FastText, TextCNN, BiLSTM and HAN). As a result, Vader Sentiment Lexicon not only works with Transformer based feature extractors, but also provides some performance enhancements to most other baseline models.

Table 6. Model internal comparison experiment.

Model	Random	Word2vec	VADER	Trans.	$MicroF_1$	$MacroF_1$
(1)	✗	✓	✓	✓	**61.89**	**57.73**
(2)	✗	✗	✓	✗	40.79	25.70
(3)	✓	✗	✗	✓	59.25	53.06
(4)	✗	✓	✗	✓	61.55	56.61
(5)	✓	✗	✓	✓	61.12	54.23

To help understand our model, we summarize the influence of internal components in the proposed model, as shown in Table 6. The experimental results show that our model performs best on $MicroF_1$ and $MacroF_1$, which proves that the model can learn the sentiment features and accurately extract the emotional information from online reviews. Model 2 reaches $MicroF_1$ and $MacroF_1$ scores by 40.79% and 25.70% with Vader sentiment analyzer, indicating that Vader sentiment analyzer played a positive role in sentiment classification. Model 3 adopts the random initialization to obtain text representation, and the model does not incorporate the sentiment vectors extracted by Vader sentiment analyzer. Model 4 uses the Vader sentiment analyzer and fuses the sentiment lexical features with the text features. It is observed that the $MicroF_1$ and $MacroF_1$ of Model 5 are increased by 1.87% and 1.17% respectively compared with Model 3, which verifies the effectiveness of Vader again. Unlike model 3, model 4 uses word2vec to train word vectors based on corpus. Comparing Model 3 with Model 4 and Model 1 with Model 5, it can be found that word vectors trained by Word2vec are more effective than randomly initialized word vectors, with an increase of about 1–4% in $MicroF_1$ and $MacroF_1$ evaluations.

4.6 Error Analysis

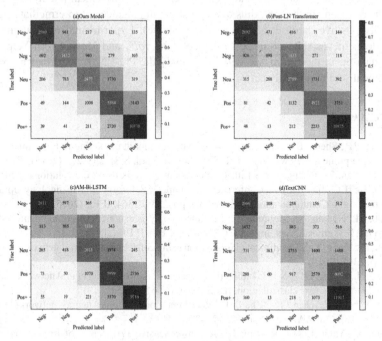

Fig. 2. Visualization of confusion matrix

We perform error analysis of the models' results on Yelp dataset and analyze category misjudgments using confusion matrix visualization. The specific results are shown in Fig. 2, where the vertical axis represents the true label and the horizontal axis represents the predicted label results. The confusion matrix in the figure contained 25 squares, in which the numbers in the diagonal squares represented the number of correct classifications and the numbers in the other squares represented the number of wrong classifications. Model (a), (b), (c) and (d) in the figure represent SPRLNT, Post-LN Transformer, AM-Bi-LSTM and TextCNN respectively.

The darker the diagonal color is in the figure, the more accurate the classification is. Among them, the diagonal color of Model (a) is the deepest, which indicates that Model (a) has better sentiment classification performance than other models. Especially, compared with Model (b), (c) and (d), Model (a) adds 734, 544 and 1210 correct classification samples in terms of the polarity of Neg respectively, which verifies the effectiveness of our model for fine-grained sentiment classification.

5 Conclusion and Future Work

In this work, we propose a model based on Pre-LN Transformer network, and integrate sentiment lexical features for improving fine-grained sentiment classification of online reviews, especially in restaurants and food. Specifically, we use the Pre-LN Transformer to extract text features, which can overcome the faults of exiting models. On this basis, we reconstruct the Vader lexicon and further earn sentiment lexical features by quantifying the emotions in the text, thus distinguishing the nuances of intermediate polarity (generally positive, neutral, and generally negative). Experimental results show that our model achieves state-of-art performance in $MicroF_1$ and $MacroF_1$ on Yelp online review dataset. In the future, we will try to apply the Pre-LN Transformer network to multi-modal sentiment analysis.

References

1. Zhang, P.P., Sheng, B., Sun, G.: Electronic word-of-mouth marketing in E-commerce based on online product reviews. Int. J. U E Serv. Sci. Technol. **8**(8), 253–262 (2015)
2. Xi, O., Pan, Z., Cheng, H.L., Liu, L.: Sentiment analysis using convolutional neural network. In: IEEE International Conference on Computer & Information Technology Ubiquitous Computing & Communications Dependable, pp. 2359–2364. IEEE (2015)
3. Santos, I., Nedjah, N., Mourelle, L.: Sentiment analysis using convolutional neural network with fastText embeddings. In: LA-CCI 2017, pp. 1–5. IEEE (2017)
4. Kim, H., Jeong, Y.-S.: Sentiment classification using convolutional neural network. Appl. Sci. **9**(11), 2347 (2019)
5. Zheng, J., Zheng, L.: A hybrid bidirectional recurrent convolutional neural network attention-based model for text classification. IEEE Access **7**, 673–685 (2019)
6. Lai, S., Xu, L., Liu, K., Zhao, J.: Recurrent convolutional neural networks for text classification. In: AAAI 2015, pp. 2267–2273 (2015)
7. Yang, X., Yang, L., Bi, R., Lin, H.: A comprehensive verification of transformer in text classification. In: Sun, M., Huang, X., Ji, H., Liu, Z., Liu, Y. (eds.) CCL 2019. LNCS (LNAI), vol. 11856, pp. 207–218. Springer, Cham (2019). https://doi.org/10.1007/978-3-030-32381-3_17
8. Bugueno, M., Mendoza, M.: Learning to combine classifiers outputs with the transformer for text classification. Intell. Data Anal. **24**(3), 15–41 (2020)

9. Xiong, R., et al.: On layer normalization in the transformer architecture. In: PMLR 2020, pp. 10524–10533 (2020)

10. Hutto, C.J., Gilbert, E.: VADER: a parsimonious rule-based model for sentiment analysis of social media text. In: AAAI 2015, pp. 8109–8122 (2015)

11. Georgios, P., Mike, T.: Twitter, MySpace, Digg: unsupervised sentiment analysis in social media. ACM Trans. Intell. Syst. Technol. 3(4), 1–19 (2012)

12. Pang, B., Lee, L., Vaithyanathan, S.: Thumbs up? Sentiment classification using machine learning techniques. In: ACL 2002, pp. 79–86 (2002)

13. Krishna, A., Akhilesh, V., Aich, A., Hegde, C.: Sentiment analysis of restaurant reviews using machine learning techniques. In: Sridhar, V., Padma, M.C., Rao, K.A.R. (eds.) Emerging Research in Electronics, Computer Science and Technology. LNEE, vol. 545, pp. 687–696. Springer, Singapore (2019). https://doi.org/10.1007/978-981-13-5802-9_60

14. Tang, D., Qin, B., Liu, T.: Deep learning for sentiment analysis: successful approaches and future challenges. Wiley Interdisc. Rev.: Data Min. Knowl. Discov. 5(6), 292–303 (2015)

15. Hameed, Z., Garcia-Zapirain, B.: Sentiment classification using a single-layered BiLSTM model. IEEE Access. 8(99), 73992–74001 (2020)

16. Vaswani, A., et al.: Attention is all you need. In: NIPS 2017, pp. 6000–6010 (2017)

17. Yang, Z., Yang, D., Dyer, C., He, X., Smola, A.: Hierarchical attention networks for document classification. In: ACL 2016., pp. 1480–1489 (2016)

18. Tang, G., Muller, M., Rios, A., Sennrich, R.: Why self-attention? A targeted evaluation of neural machine translation architectures. In: EMNLP 2018. pp. 4263–4272 (2018)

19. Vaswani, A., et al.: Tensor2Tensor for neural machine translation. In: Proceedings of the 13th Conference of the Association for Machine Translation in the Americas, pp. 193–199 (2018)

20. Wang, Q., Li, B., Xiao, T., et al.: Learning deep transformer models for machine translation. In: ACL 2019, pp. 1810–1822 (2019)

21. McAuley, J.J., Leskovec, J.: From amateurs to connoisseurs: modeling the evolution of user expertise through online reviews. In: Proceedings of the 22nd International Conference on World Wide Web, pp. 897–907 (2013)

22. Kim, Y.: Convolutional neural networks for sentence classification. In: EMNLP 2014, pp. 1746–1751 (2014)

23. Joulin, A., Grave, E., Bojanowski, P., Mikolov. T.: Bag of tricks for efficient text classification. In: ACL 2017, pp. 427–431 (2017)

24. Schuster, M., Paliwal, K.K.: Bidirectional recurrent neural networks. IEEE Trans. Signal Process. 45(11), 2673–2681 (1997)

25. Wang, M., et al.: Sentiment analysis based on attention mechanisms and bi-directional LSTM fusion model. In: SmartWorld/SCALCOM/UIC/ATC/CBDCom/IOP/SCI 2019, pp. 865–868. IEEE (2019)

Adversarial Context-Aware Representation Learning of Multiword Expressions

Bo An[1,2(✉)]

[1] Institute of Ethnology and Anthropology, Chinese Academy of Social Sciences, Beijing, China
[2] Institute of Software, Chinese Academy of Sciences, Beijing, China

Abstract. Word composition is a promising method to learn the representations of long text. Unfortunately, the representations of non-compositional multiword expressions (e.g., go banana) can not be inferred by word composition. Most current methods regard a multiword expression as a single word, and learn its representation in the same way as word representations. However, many multiword expressions are ambiguous, that they express distinct meanings (literal or idiomatic) in different contexts. To resolve this problem, this paper proposes an adversarial context-aware representation learning method for multiword expressions, which generates representations based on the contexts of their occurrences. An adversarial training framework is introduced for further enhancing the representation learning method. The experimental results verify the beneficial of sense disambiguation of multiword expression for representations learning, and the proposed method achieved competitive performances on both the idiom token classification and compositionality prediction tasks.

Keywords: Multiword expression · Adversarial learning · Context aware

1 Introduction

Multiword expression (MWEs) is fixed collocations made up of a sequence of two or more words that exhibit some kind of non-compositional. In a word, the meanings of a MWE is not predictable from the composition of its constituent words. For instance, '*kick the bucket, go banana*'. The number of MWEs is of the same order of magnitude as the number of single words, there are about 41% of entries are multiword expressions in WordNet 1.7 (Felbaum 1998), as a result the MWEs are prevalent in text. So it is critical to appropriately handle the MWEs in many natural language processing (NLP) tasks, like machine translation [1], natural language understanding [2] and paraphrase detection [3].

Recently, the distributional representations of text have achieved significant success in many NLP tasks, such as machine translation, question answering and natural language understanding[1, 2]. The distributional representation learning

H. Lin et al. (Eds.): CCIR 2021, LNCS 13026, pp. 112–126, 2021.
https://doi.org/10.1007/978-3-030-88189-4_9

methods mainly based the distributional hypothesis (Harris, 1954), i.e., '*words in similar contexts tend to have similar meanings*'. And word composition is a promising approach to learn the representations of long text, such as phrase and sentence. The word composition models are mainly based on the principle of compositionality, i.e., '*The meaning of a complex determined by the meanings of its constituent expressions and the rules used to combine them*' [4].

It is obvious that, the meanings of multiword expressions violate the principle of word compositionality, and the proper representations of MWEs can not be generated by word composition method. As a result, the way of handling multiword expressions is a meaningful research topic. Compositionality prediction is one of the important tasks for processing MWEs, which aims to determine the degree to which the meaning of the parts of a MWE combine (literal meaning) to predict the meaning (idiomatic meaning) of the whole. Currently, the mainstream approach of representing the MWEs is to treat an occurrence of multiword expressions as a single word, and learns its representation in the same way of learning word representations.

Most of these methods regard MWEs as unambiguous, unfortunately, a multiword expression may express distinct meanings in various contexts. For instance, '*kick the bucket*' expresses its literal meaning and idiomatic meaning in '*The old man finally kicks the bucket at ninety five.*' and in '*Ryan runs to kick the bucket in agitation.*', respectively. Such a uniform representation schema inevitably introduces a lot of noisy contexts for learning representations of multiword expressions.

On the other hand, the idiom token classification aims at distinguishing the literal meaning from the idiomatic meaning of an occurrence of a MWE. Currently, most of the methods focus on finding meaningful features and proper classifiers for this task [5,6]. Unfortunately, this line of works mainly concern on disambiguating the occurrences of MWEs, which is difficult to be utilized by the subsequence NLP technologies directly.

To address the issues, this paper proposes a context-aware multiword expressions representation learning model, which detects the idiomatic occurrences of multiword expressions based on their contexts and learns the representations (literal and idiomatic) for MWEs. The proposed method firstly generates the representation of an occurrence of a multiword expression, and disambiguates the occurrence based on the generated representation and its literal representation. Specifically, if the literal representation more accordant with the context of a specific occurrence, the more likely the occurrence of the MWE expresses its literal meaning. Secondly, all of the idiomatic instances of a multiword expression are summarized to generate the representation of idiomatic meaning of the multiword expression, which is employed to predict its compositionality. For that purpose, our work proposes a generator network, which predicts the representation of a multiword expression depended on its context. Inspired by the generative adversarial network (GAN) [7], we present an adversarial training framework which introduces a discriminative model to further enhance the generator by providing gradient information. Specifically, the purpose of the discriminator is to correctly distinguish the representation of target multiword expression from the generated vector. And the generator is intended to confuse the discriminator by generating vector as in accordance with the context as possible. And the discriminator and generator play a minimax game to enhance each other. In this way, we can train a generative model which generates the proper vector for any given context.

The main contributions of this work are twofold: (1) To the best of our knowledge, this is the first work which takes the ambiguity of multiword expressions into consideration for representation learning of multiword expressions. (2) This paper proposes an adversarial framework to enhance the generated representations.

2   Related Work

This section briefly reviews related work, including distributional representation learning, generative adversarial network, idiom token classification and multiword expression compositionality prediction.

Distributional representation of text methods have demonstrated their utility in a wide range of NLP tasks. Several well-known models include C&W [8], word2vec [9], GloVe [10] and AutoExtend [11] for learning word embeddings. There are some models which take the ambiguity of words into consideration, Multi-Prototype model [12] and MSSG [13] learns multiple vectors for each word in the vocabulary. TWE [14] incorporates the topic labels to disambiguate the meanings of a word in different contexts. Many models have been developed for word composition to represent long text, e.g., simple models such as element-wise addition, element-wise multiplication [9,15,16] and neural network based models such as recurrent neural network [17], gated recurrent neural network, Long-Short Term Memory network [18] and convolution neural network [19].

Adversarial networks [7] (GAN) have recently surfaced as a general tool of measuring equivalence between distributions and it has proven to be effective in a variety of tasks [20,21]. The adversarial framework proposes a discriminative model to enhance the generative model by playing a minimax game between them.

Idiom token classification is a task of deciding whether each occurrence of a compound expresses literal or idiomatic meanings. The structure and distribution information are utilized in traditional methods [22,23]. More recently, PCA and LDA are utilized for the task [5,24]. [25] assumes that the literal meanings of a multiword expression fit the context better than the idiomatic ones, and utilizes inner production of vectors of context and composed representation to classify the meanings of MWEs. [6] employs Skip-Thought Vectors to incorporate the context information beyond the target sentences for classification, which achieves current SOA performances.

The early methods on multiword expressions compositionality prediction task focus on utilizing statistical features (e.g., the features for word sense disambiguation) to detect compositionality of a variety of MWEs [26–31]. [32] presents a multilingual translation-based method to detection compositionality. Recent works employ distributional representation for compositionality prediction task.

[33] utilizes the non-substitutability property of MWEs to predict the compositionality of multiword expressions based on distributional representations. [34] attempts to utilize word embeddings to predict compositionality of multiword expressions. [35] models the compositionality of multiword expressions as the degree to which of the a multiword expression can be modelled by the learned semantic composition function, the less of a MWE can be modelled by the learned composition function, the smaller compositionality of the multiword expression. [36] presents an adaptive joint learning method for compositionality detection and embeddings learning, but this method only applied for Verb-Object type of MWEs. [37] gives a detail evaluation of different methods and the level of corpus preprocessing for this task, and achieves the state-of-the-art performances on this task.

By contrast, this paper aims to exclude noisy context for learning idiomatic representations of multiword expressions, and to improve the performances of following tasks based on the representations.

3   Context-Aware Multiword Expressions Representation Learning

In this section, we present our adversarial context-aware multiword expression representation learning model. We first describe the method of learning the representations for literal meanings of multiword expressions, and then we propose an adversarial framework to train the generator which generates representation of a MWE occurrence based on its context. Secondly, the model disambiguates all the occurrences of MWEs, and learns the representations of their idiomatic meanings. Finally, the method of predicting compositionality of multiword expressions is detailed described.

3.1   Literal Meaning Representation

The literal meaning of a multiword expression obedient to the principle of compositionality, which can be obtained by composing its constituent words. There are a lot of works devoted to word composition, e.g., simple models such as element-wise addition, element-wise multiplication [9,15,16] and neural network based models such as recurrent neural network [17], gated recurrent neural network, Long-Short Term Memory network [18] and convolution neural network [19]. Because most of multiword expressions with limited lengths, usually contain two or three words, this paper employs the simple normalized element-wise addition as the composition model to infer the literal representations, which has proven to be both robust and effective in many tasks. Another advantage of this simple model is that it enables more equitably compared with other baseline models. Concretely, the literal representation of a multiword expression is derived as follows:

$$m\vec{w}e_L = \sum_{i=1}^{n} \frac{\vec{w_i}}{|\vec{w_i}|} \tag{1}$$

where n is the number of words in the MWE, $\vec{w_i}$ is the representation of pth word, and $|\vec{w}|$ is the norm of the word vector.

3.2   Adversarial Context-Aware Representation Learning

The context-aware representation learning aims at learning representation of a multiword expression based on the context of their occurrences. In other words, the model predicts the representation of target text based on its context. Based on this comprehension, given the context, the model should generate the representation of a target text with random length, including word and phrase. Since there is little corpus which labels each occurrence of multiword expressions with idiomatic meaning or literal meaning, and labelling a new corpus is laborious. We can train our generative model based on (context, word) tuples, which capable generate proper representations of a multiword expression based on its context. In this way, each of the words in text corpus and its context are used as training data for our model, which can be easily constructed automatically from text corpus. Specifically, given a sentence $S = [w_1, ..., w_{i-1}, w_i, w_{i+1}, ..., w_k]$, training a θ-parameterized generative model G_θ to predict a proper vector representation of $\vec{w_i}$ based on its context $C = [w_1, ..., w_{i-1}, w_{i+1}, ..., w_k]$.

Based on the above dataset, the trained generator G_θ is capable to produce a proper representation for a multiword expression or a word, which accordant with its context. In addition, we also introduce a ϕ-parameterized discriminative model D_ϕ to provide guidances for enhancing the generator G_θ. For a given target word/MWE and its context, the problem can be regarded as conditional probability $p_{true}(w|c)$, which depicts the true distribution over the candidate vocabulary with respect to the context. Given a text corpus, each word and its context in a sentence could be used as training data pair $(context, word)$ for constructing representation generative models.

Generative Model. The generative model D_ϕ tries to generate accordant vector for the given context. In other words, it aims at approximating the true relevance word as $p_{true}(w|c)$ as much as possible. Specifically, this paper first map the discrete words in context into distributional embedding vectors using lookup embedding matrix for each word in the vocabulary. Because the target word in different position of the same context may express distinct meanings, so we take the position information with respect to target word into consideration as following:

$$\vec{v_i} = \vec{w_i} \oplus \vec{p_i} \tag{2}$$

where $\vec{w_i} \in d^m$ is the representation of w_i, the $\oplus$ is a general concatenation operation, which concatenates these representations into a single vector, $p_i \in d^n$ is the relative position of w_i, and $\vec{v_i} \in d^{m+n}$ is the input vector for the following layer. Specifically, given a sentence $S = [w_1, ..., w_{i-1}, w_i, w_{i+1}, ..., w_k]$ and target word w_i, the position of w_{i-1} is -1, and the position of w_k is $k - i$.

To generate representation for a context, we introduce a bidirectional LSTM (LSTM) [38] network with an average pooling layer, and then two fully-connected feedforward layers followed close behind to produce accordant vector for the context.

Discriminative Model. The discriminative model $f_\phi(w, c)$ which, in contrary, aiming at discriminating the true data $(context, \vec{w}_i)$ from the ill-matched ones, where the goodness of matching generated by $f_\phi(w, c)$ depends on the accordance of $\vec{w}_i$ to the *context*. In this paper, we compute the accordance between context and representation vector of the target word instead of the word itself, because our goal is to generate the appropriate vector of target word. Another benefit of the objective function is that it can be directly optimized by gradient descent as the original generative adversarial network formulation. Concretely, the same as the generative model, the discriminative model D_ϕ first converts the words in context into vectors as Formula (1), and a BLSTM layer is utilized to encode the context following with a average pooling layer. And then two fully-connected feedforward layers followed close behind to produce the representation of context. Finally, the vectors of context and target word are concatenated into a single vector, which used as the input of a softmax layer to infer the accordance of (context, word) tuple.

Algorithm 1. Adversarial Representation Learning algorithm

Require: generator G_θ; discriminator D_ϕ;   training dataset: T = (context, word) ;
1: Initialise G_θ and D_ϕ with random weights θ, ϕ;
2: Pre-train G_θ and D_ϕ using T;
3: **repeat**
4:   **for** g-steps do **do**
5:     Sample minibatch of m tuples from T;
6:     G_θ generates a vector for each context in T;
7:     Update generator parameters via descending its stochastic gradient;

$$\nabla_\phi \frac{1}{m} \sum_{i=1}^{m} log(1 - D(G_\theta(c_i)|c_i))$$

8:   **for** d-steps do **do**
9:     Sample minibatch of m tuples from T;
10:    Use current G_θ to generate a negative vector for each tuple in the above minibatch;
11:    Update the discriminator by ascending its stochastic gradient;

$$\nabla_\phi \frac{1}{m} \sum_{i=1}^{m} [log D(\vec{w}_i|c_i) + log(1 - D(G_\theta(c_i)|c_i))]$$

12: **until** converges

A Minimax Framework. Inspired by the idea of generative adversarial network, this paper tries to unify these two different types of models by letting them play a minimax game: the discriminative model D_ϕ would try to distinguish between the true target word representation and the generated vector predicted by its opponent model, where the generative model G_θ aims at fooling the discriminator model by generating vector as same as the representation of target word. Formally, we have:

$$J^{G,D} = \min_{\theta} \max_{\phi} \sum_{n=1}^{N} (E_{\vec{w} \sim p_{data}(\vec{w}|c)}[logD(\vec{w}|c)]$$
$$+ E_{\vec{w} \sim p_{\theta}(\vec{w}|c)}[log(1 - D(\vec{w}|c))]) \tag{3}$$

The overall logic of our proposed adversarial representation learning framework is summarized in Algorithm 1. Before the adversarial training, the generator and discriminator can be initialised by their conventional models. In this way, the generative model G_{θ} can be optimized not only by the training tuples, but also by the gradient information from discriminative model D_{ϕ}.

3.3   Idiomatic Token Disambiguation

In this section, we first introduce the method of disambiguating occurrences of a multiword expression based on the representation generated by the generative model G_{θ} based on their contexts. Secondly, we infer the accordance of a generated vector for a specific context by calculating the cosine similarity between the generated vectors and literal representation of the multiword expression. And then the KNN algorithm is employed to cluster all of the accordances of a specific multiword expression into two categories (k is set to 2). Finally, we calculate the center score for one of the clusters as following:

$$score_{center} = \frac{1}{M} \sum_{i=1}^{M} accordance_i \tag{4}$$

where M refers to the number of accordances in the clusters.

We identify the idiomatic meaning of an occurrence of a multiword expression based on an additional assumption that: the accordance reflects the degree to which the literal meaning of a multiword expression fits with the context of an occurrence, and the occurrences with smaller accordances express idiomatic meanings of the multiword expression. All the occurrences in the cluster with a smaller center score are labelled as idiomatic for a specific multiword expression, and the rest of its occurrences are regarded as literal.

3.4   Compositionality Prediction

In this section, we first present the method of generating idiomatic representation of a multiword expression based on its idiomatic occurrences and their generated representations. The idiomatic representation of a multiword expression is inferred as following:

$$\vec{mwe}_I = \frac{1}{n} \sum_{i=1}^{n} \vec{I}_i \tag{5}$$

where n refers to the number of idiomatic occurrences of the multiword expression, $\vec{I_i}$ is the generated vectors of the ith idiomatic occurrence. And then the compositionality of the multiword expression is computed as:

$$compositionality_{mwe} = \frac{m\vec{w}e_I \cdot m\vec{w}e_L}{|m\vec{w}e_I|_2 |m\vec{w}e_L|_2} \tag{6}$$

where $m\vec{w}e_I$ and $m\vec{w}e_L$ are the idiomatic and literal representations of the MWE and the $\cdot$ represents the dot product operator.

4 Experiments

In this section, we first describe the settings in our experiments[3], and then we conduct experiments of idiom token classification and multiword expressions compositionality prediction tasks and compare our models with strong baselines.

4.1 Evaluation Settings

Text Corpus. We use the lemmatized and POS-tagged versions of the ukWaC for English (about 2 billion tokens)[1]. The text corpus is preprocessed by removing stopwords, lemmatization, lowercasing and replacing numbers by a label.

Settings. The word embeddings is pre-trained by the word2vec tool's CBOW algorithm[2], with the dimension of word vector is 200, the windows size is 5, the number of iterations is 5, 10 negative samples and the number of iterations is 3. The initial values for a weight matrix were uniformly sampled from the symmetric interval $[-\frac{1}{\sqrt{n}}, \frac{1}{\sqrt{n}}]$, where n is the dimension of the embeddings. The dimension of the position embedding is set as 100. The stochastic gradient descent (SGD) is employed to minimize of the objective function of the neural network. The number of BLSTM units is set as 300, the forward and reverse running LSTM-networks had the same number of recurrent units. The mini-batch size is 100, the dropout rate is 0.4 and the learning rate of SGD is 0.01. The number of units of the fully-connected hidden layer is 400, and output of the full-connected layer is a vector of 200 dimensions. We implement our adversarial framework using Tensorflow deep learning framework[3]. CARL refers to the proposed model which is trained based on the (context, word) tuples, and the model which enhanced by the adversarial training framework is represented as CARL-GAN.

To evaluate the effectiveness of the proposed models, for each of the multiword expression dataset, we firstly replace all the occurrences of multiword expressions with the concatenated of their constituent words. For example, all the occurrences of *end user* are replaced by *end_user*. Secondly, the pre-processed text corpus is utilized to learn word embeddings, and the learned embeddings for the multiword expressions as the base idiomatic representations.

---

[1] http://fraublucher.sslmit.unibo.it/wac/.
[2] https://code.google.com/archive/p/word2vec/.
[3] https://www.tensorflow.org.

4.2   Idiom Token Classification

Idiom token classification is a task of deciding for a set of potentially idiomatic multiword expressions whether each occurrence of a MWE is a literal or idiomatic usage of the phrase. In this paper, we employ this task to evaluate the effectiveness of the representations of MWEs from our model for identifying idiomatic expressions.

Dataset. To compare with the baseline systems, we employ VNC-Token dataset [22] to conduct experiments in this task. There are 53 different verb-noun constructions (VNCs) extracted from British National Corpus (BNC). There are 2984 sentences contain one of the VNCs, which are labelled as I (idiomatic); L (literal); or Q (unknown). To fairly compare with the baselines, 28 of the VNCs which have a reasonably balanced representation (with similar numbers of idiomatic and literal occurrences in the corpus) are used in this experiment, and removing the sentence with Q label. The precision, recall and F1-score are employed to evaluate models.

Following [5,6], we conduct two experiments for this task. The purpose of the former one is to detailed evaluate different models on four multiword expressions (BlowWhistle, LoseHead, MakeScene and TakeHeart). To make a fair comparison with the baseline methods, we construct the training and test datasets for each of the multiword expressions by randomly sampling MWEs following the same distribution as [5]. And the data statistics are listed in Table 1.

Table 1. The sizes of the samples for each expression and the split into training and test sets. The numbers in parentheses indicates the number of idiomatic labels within the set.

Expression	Samples	Train size	Test size
BlowWhistle	78 (27)	40 (20)	38 (7)
LoseHead	40 (21)	30 (15)	10 (6)
MakeScene	50 (30)	30 (15)	20 (15)
TakeHeart	81 (61)	30 (15)	51 (46)

We compare our models with two strong baselines: [5] and Sen2vec [6], the Sen2vec had achieved competitive performances on this task. In addition, we implement a base model (CBOW) with the pre-trained word embeddings, which classifies the meanings of an occurrence based on the cosine similarity of the context vector and idiomatic representation of the MWE, which is directly inferred from the lookup embedding matrix.

The conclusions can be drawn from Table 2 are: (1) The proposed models achieves significant improvements over the base CBOW model, which verifies the beneficial of avoiding noisy context. (2) The adversarial training method consistently enhances the performances of idiom token classification. (3) The

Table 2. Results in terms of precision (P.), recall (R.) and f1-score (F1) on the four chosen expressions. The bold values indicates the best results for that expression in terms of f1-score.

Models	BlowWhistle			LoseHead			MakeScene			TakeHeart		
	P.	R.	F1	P.	R.	F1	P.	R.	F1	P.	R.	F1
Peng et al. (2014)												
FDA-Topics	0.62	0.60	0.61	0.76	0.97	0.85	0.79	0.95	0.86	0.93	0.99	**0.96**
FDA-Topics+A	0.47	0.44	0.45	0.74	0.93	0.82	0.82	0.69	0.75	0.92	0.98	0.95
FDA-Text	0.65	0.43	0.52	0.72	0.73	0.72	0.79	0.95	0.86	0.46	0.40	0.43
FDA-Text+A	0.45	0.49	0.47	0.67	0.88	0.76	0.80	0.99	0.88	0.47	0.29	0.36
SVMs-Topics	0.07	0.40	0.12	0.60	0.83	0.70	0.46	0.57	0.51	0.90	1.00	0.95
SVMs-Topics+A	0.21	0.54	0.30	0.66	0.77	0.71	0.42	0.29	0.34	0.91	1.00	0.95
SVMs-Text	0.17	0.90	0.29	0.30	0.50	0.38	0.10	0.01	0.02	0.65	0.21	0.32
SVMs-Text+A	0.24	0.87	0.38	0.66	0.85	0.74	0.07	0.01	0.02	0.74	0.13	0.22
Sen2vec												
KNN-2	0.61	0.41	0.49	0.30	0.64	0.41	0.55	0.89	0.68	0.46	0.96	0.62
KNN-3	0.84	0.32	0.46	0.58	0.65	0.61	0.88	0.88	0.88	0.72	0.94	0.81
KNN-5	0.79	0.28	0.41	0.57	0.65	0.61	0.87	0.83	0.85	0.73	0.94	0.82
KNN-10	0.83	0.30	0.44	0.28	0.68	0.40	0.85	0.83	0.84	0.78	0.94	0.85
Linear SVM	0.77	0.50	0.60	0.72	0.84	0.77	0.81	0.91	0.86	0.73	0.96	0.83
Grid SVM	0.80	0.51	0.62	0.83	0.89	0.85	0.80	0.91	0.85	0.72	0.96	0.82
SGD SVM	0.70	0.40	0.51	0.73	0.79	0.76	0.85	0.91	0.88	0.61	0.95	0.74
Ours												
CBOW	0.7	0.39	0.50	0.72	0.69	0.70	0.72	0.7	0.71	0.58	0.83	0.68
CARL	0.79	0.46	0.58	0.81	0.79	0.80	0.84	0.85	0.84	0.72	0.95	0.82
CARL-GAN	0.84	0.53	**0.65**	0.87	0.86	**0.86**	0.87	0.91	**0.89**	0.77	0.96	0.85

proposed CARL-GAN model had achieved comparable performances with state-of-the-art results on the task.

In the latter experiment of this task, we conduct idiom token classification on all of 28 MWEs. For fairly compared with Sen2vec, this paper constructs the datasets following the same way and distributions as [6]. And the overall results are listed in Table 3.

From Table 3 we can see that: (1) Both the CARL and CARL-GAN significantly improved the base CBOW models, which verifies that the learned representations of multiword expressions by excluding the noisy context for this task. (2) The CARL-GAN had achieved the best recall and f1-score on this dataset.

In summary, the proposed method can substantially enhance the representations of MWEs for idiomatic token classification task. The gradient information from the discriminative model is beneficial for improving the generative model. It is important to detect the idiomatic occurrences of MWEs for representation learning.

Table 3. Results in terms of precision (P.), recall (R.) and f1-score (F1) on all of the 28 MWEs from different models.

Models		P.	R.	F1
Sen2vec	Linear-SVM-GE	**0.84**	0.80	0.83
	Grid-SVM-GE	**0.84**	0.80	0.83
	SGD-SVM-GE	0.79	0.79	0.78
Ours	CBOW	0.71	0.74	0.72
	CARL	0.82	0.81	0.81
	CARL-GAN	0.86	**0.83**	**0.84**

4.3 Compositionality Prediction

To evaluate our models on compositionality prediction task, we conduct experiments on three widely-used datasets for English: Reddy, Reddy++ and Farahmand. The Spearman ρ correlation between the ranking from golden data and predictions from models is used to assess different models.

Datasets. Reddy is a widely used dataset collected with Mechanical Turk [31], which contains compositional scores for 90 multiword expressions and each of their constituent words. Most of the compound are formed by nouns, such as *end user*. Reddy++ is a recently created dataset for compositionality prediction task [39]. It extends the Reddy dataset to 180 entries, and introduces some adjective-noun compounds. The compositional score for each compound in both Reddy and Reddy++ is calculated by averaging the manually annotated scores assigned to the entire expression.

Farahmand [40] is a relatively large dataset, which contain 1042 multiword expressions extracted from Wikipedia with binary non-compositional judgements from four experts for each compound. In our experiments, the average of the judgements is used as the compositonality of a compound.

This paper first calculates Spearman ρ correlation on the Reddy and Reddy++ datasets. We compare our method with several strong baselines: the weighted addition and multiplication methods [31], the multi-way translation-based method [32], a word embeddings based method from [34] and PMI, Word2Vec and Glove methods [37]. In addition, the pre-trained word embeddings (CBOW) for our model is utilized as another baseline, and the compositionlaity is measured by the cosine similarity between literal representation and idiomatic representation of a MWE. To the best of our knowledge, [37] had achieved the current state-of-the-art performances on these datasets, that achieves significant improvements over the other models, and we list their best results for each dataset. The overall results are presented in Table 4.

Table 4. The Spearman's correlation ρ of different methods on Reddy, Reddy++ and Farahmand datasets.

Models	Reddy	Reddy++
Addition	0.71	0.63
Multiplication	0.65	0.57
Translation	0.74	–
[34]	0.75	–
Best w2v	0.82	**0.73**
Best PPMI	0.80	0.72
Best glove	0.76	0.66
CBOW	0.72	0.63
CARL	0.78	0.69
CARL-GAN	**0.83**	**0.73**

From Table 4, we can infer that: (1) Our CARL-GAN model had achieved the best performances on both of the datasets. (2) Both the CARL and CARL-GAN models significantly improved the CBOW based on pre-trained word embeddings, which verifies the beneficial of multiword expression disambiguation. (3) The adversarial training method significantly improves our CARL model by 6.4 and 5.8% on Reddy and Reddy++ respectively, that demonstrates the effectiveness of adversarial training framework. (4) Our model had similar performances with [37], but they achieved their best performances by distinct parameters on different datasets and different methods of pre-processing text corpus. And the results may be optimized based on their word embeddings, but that is not the focus of this article.

Following [35], we evaluate our models using F1-score (BF1) on Farahmand dataset. In this experiment, we compare our models with [35,37] and the pre-trained word embeddings using CBOW for our model. [37] had achieved current state-of-the-art results on this dataset. The overall results are shown in Table 5.

Table 5. The BF1 scores on Farahmand dataset.

Models	Farahmand
[35]	0.49
Best w2v	0.51
BestPPMI	0.52
Best glove	0.40
CBOW	0.43
CARL	0.48
CARL-GAN	**0.53**

We can infer from Table 5 that: both of CARL and CARL-GAN models significantly improve the performances achieved by the pre-trained word embeddings, and CARL-GAN model achieves the best BF1 score on Farahmand.

In a word, our model had achieved comparable performances as state-of-the-art methods, and even better results on some Reddy and Farahmand datasets. Both the CARL and CARL-GAN significantly improved the performances of pre-trained word embeddings baseline, which verifies the beneficial of detecting idiomatic meanings of a multiword expression for representation learning. The adversarial training framework consistently enhances the representations from the generative model through the gradient information from its competing discriminative model.

5   Conclusions

In this paper, we propose an adversarial context-aware representations learning model to learn the embeddings of multiword expressions. The model utilizes the context to predict the representation vector of occurrences of MWEs, which are utilized to detect idiomatic occurrences of MWEs for learning the idiomatic representations of multiword expressions. Experimental results verified the effectiveness of the generated representations of MWEs on both the idiom token classification and compositionality prediction tasks. The adversarial training framework enhances the performances by optimizing the generative model by providing additional gradient information.

Acknowledgments. This work is supported by the National Natural Science Foundation of China under Grants no. 62076233 and the Major innovation project of Chinese Academy of Social Sciences no. 2020YZDZX01-2.

References

1. Zou, W.Y., Socher, R., Cer, D.M., Manning, C.D.: Bilingual word embeddings for phrase-based machine translation. In: EMNLP, pp. 1393–1398 (2013)
2. Yang, X., et al.: End-to-end joint learning of natural language understanding and dialogue manager. In: IEEE International Conference on Acoustics, Speech and Signal Processing, pp. 5690–5694 (2017)
3. Cheng, J., Kartsaklis, D.: Syntax-aware multi-sense word embeddings for deep compositional models of meaning. arXiv preprint arXiv:1508.02354 (2015)
4. Pelletier, F.J.: Did Frege believe Frege's principle? J. Logic Lang. Inform. **10**(1), 87–114 (2001)
5. Peng, J., Feldman, A., Vylomova, E.: Classifying idiomatic and literal expressions using topic models and intensity of emotions. In: Empirical Methods in Natural Language Processing, pp. 2019–2027 (2014)
6. Salton, G., Kelleher, J.D., Ross, R.J.: Idiom token classification using sentential distributed semantics. In: Meeting of the Association for Computational Linguistics (2016)
7. Goodfellow, I.J., et al.: Generative adversarial networks. Adv. Neural. Inf. Process. Syst. **3**, 2672–2680 (2014)

8. Collobert, R., Weston, J.: A unified architecture for natural language processing: deep neural networks with multitask learning. In: Proceedings of the 25th International Conference on Machine Learning, pp. 160–167. ACM (2008)
9. Mikolov, T., Sutskever, I., Chen, K., Corrado, G.S., Dean, J.: Distributed representations of words and phrases and their compositionality. In: Advances in Neural Information Processing Systems, pp. 3111–3119 (2013)
10. Pennington, J., Socher, R., Manning, C.D.: GloVe: global vectors for word representation. In: EMNLP, vol. 14, pp. 1532–1543 (2014)
11. Rothe, S., Schütze, H.: AutoExtend: extending word embeddings to embeddings for synsets and lexemes. arXiv preprint arXiv:1507.01127 (2015)
12. Huang, E.H., Socher, R., Manning, C.D., Ng, A.Y.: Improving word representations via global context and multiple word prototypes. In: Proceedings of the 50th Annual Meeting of the Association for Computational Linguistics: Long Papers, vol. 1, pp. 873–882. Association for Computational Linguistics (2012)
13. Neelakantan, A., Shankar, J., Passos, A., McCallum, A.: Efficient non-parametric estimation of multiple embeddings per word in vector space. arXiv preprint arXiv:1504.06654 (2015)
14. Liu, Y., Liu, Z., Chua, T.S., Sun, M.: Topical word embeddings. In: AAAI, pp. 2418–2424 (2015)
15. Mitchell, J., Lapata, M.: Composition in distributional models of semantics. Cogn. Sci. **34**(8), 1388–1429 (2010)
16. Blacoe, W., Lapata, M.: A comparison of vector-based representations for semantic composition. In: Proceedings of the 2012 Joint Conference on Empirical Methods in Natural Language Processing and Computational Natural Language Learning, pp. 546–556. Association for Computational Linguistics (2012)
17. Paulus, R., Socher, R., Manning, C.D.: Global belief recursive neural networks. In: Advances in Neural Information Processing Systems, pp. 2888–2896 (2014)
18. Le, P., Zuidema, W.: Compositional distributional semantics with long short term memory. arXiv preprint arXiv:1503.02510 (2015)
19. Kalchbrenner, N., Grefenstette, E., Blunsom, P.: A convolutional neural network for modelling sentences. Eprint Arxiv **1** (2014)
20. Wang, J., et al.: IRGAN: a minimax game for unifying generative and discriminative information retrieval models (2017)
21. Liu, P., Qiu, X., Huang, X.: Adversarial multi-task learning for text classification. In: Meeting of the Association for Computational Linguistics, pp. 1–10 (2017)
22. Cook, P., Fazly, A., Stevenson, S.: The VNC-tokens dataset. In: Proceedings of the MWE Workshop ACL (2008)
23. Li, L., Sporleder, C.: Using gaussian mixture models to detect figurative language in context. In: Human Language Technologies: The 2010 Conference of the North American Chapter of the Association for Computational Linguistics, pp. 297–300 (2010)
24. Feldman, A., Peng, J.: Automatic detection of idiomatic clauses. In: Gelbukh, A. (ed.) CICLing 2013. LNCS, vol. 7816, pp. 435–446. Springer, Heidelberg (2013). https://doi.org/10.1007/978-3-642-37247-6_35
25. Peng, J., Feldman, A., Jazmati, H.: Classifying idiomatic and literal expressions using vector space representations. In: Ranlp (2015)
26. Lin, D.: Automatic identification of non-compositional phrases. In: Meeting of the Association for Computational Linguistics on Computational Linguistics, pp. 317–324 (1999)

27. Mccarthy, D., Keller, B., Carroll, J.: Detecting a continuum of compositionality in phrasal verbs. In: ACL 2003 Workshop on Multiword Expressions: Analysis, Acquisition and Treatment, pp. 73–80 (2003)
28. Venkatapathy, S., Joshi, A.K.: Measuring the relative compositionality of verb-noun (V-N) collocations by integrating features. In: Conference on Human Language Technology and Empirical Methods in Natural Language Processing, pp. 899–906 (2005)
29. Mccarthy, D., Venkatapathy, S., Joshi, A.K.: Detecting compositionality of verb-object combinations using selectional preferences. In: EMNLP-CoNLL 2007, Proceedings of the 2007 Joint Conference on Empirical Methods in Natural Language Processing and Computational Natural Language Learning, 28–30 June 2007, Prague, Czech Republic, pp. 369–379 (2007)
30. Lapata, M., Mitchell, J.: Vector-based models of semantic composition (2008)
31. Reddy, S., Mccarthy, D., Manandhar, S.: An empirical study on compositionality in compound nouns (2011)
32. Salehi, B., Cook, P., Baldwin, T.: Using distributional similarity of multi-way translations to predict multiword expression compositionality. In: Eacl (2014)
33. Farahmand, M., Henderson, J.: Modeling the non-substitutability of multiword expressions with distributional semantics and a log-linear model. In: The Workshop on Multiword Expressions, pp. 61–66 (2016)
34. Salehi, B., Cook, P., Baldwin, T.: A word embedding approach to predicting the compositionality of multiword expressions. In: Conference of the North American Chapter of the Association for Computational Linguistics: Human Language Technologies, pp. 977–983 (2015)
35. Yazdani, M., Farahmand, M., Henderson, J.: Learning semantic composition to detect non-compositionality of multiword expressions. In: Conference on Empirical Methods in Natural Language Processing, pp. 1733–1742 (2015)
36. Hashimoto, K., Tsuruoka, Y.: Adaptive joint learning of compositional and non-compositional phrase embeddings, pp. 205–215 (2016)
37. Cordeiro, S., Ramisch, C., Idiart, M., Villavicencio, A.: Predicting the compositionality of nominal compounds: Giving word embeddings a hard time. In: Meeting of the Association for Computational Linguistics, pp. 1986–1997 (2016)
38. Hochreiter, S., Schmidhuber, J.: Long short-term memory. Neural Comput. 9(8), 1735–1780 (1997)
39. Ramisch, C., Cordeiro, S., Zilio, L., Idiart, M., Villavicencio, A.: How naked is the naked truth? a multilingual lexicon of nominal compound compositionality. In: Meeting of the Association for Computational Linguistics, pp. 156–161 (2016)
40. Farahmand, M., Smith, A., Nivre, J.: A multiword expression data set: annotating non-compositionality and conventionalization for English noun compounds. In: The Workshop on Multiword Expressions, pp. 29–33 (2015)

IR in Education

Research on the Evaluation Words Recognition in Scholarly Papers' Peer Review Texts

Kun Ding[1] , Xinhang Zhao[1] , Liang Yang[2] , Kaiqiao Wang[1] ,
and Yuan Lin[1(✉)]

[1] WISELab, Dalian University of Technology, Dalian, China
{dingk,liang,zhlin}@dlut.edu.cn, {xinhang,wkqiao}@mail.dlut.edu.cn
[2] Faculty of Electronic Information and Electrical Engineering,
Dalian University of Technology, Dalian, China

Abstract. Peer review is an important means of academic evaluation. The evaluation words in peer review texts reflect the important viewpoints of reviewers. In this paper, An evaluation words recognition method for peer review texts (TransPeerBCC) is proposed based on transfer learning method and BiLSTM-CNN-CRF framework. TransPeerBCC first classifies direct evaluation word and indirect evaluation word, and then uses BiLSTM-CNN-CRF framework to identify these words. At the same time, in order to improve the recognition accuracy, the model parameters of public domain data are transferred to the peer review texts' using the transfer learning method. The effectiveness of the method is verified by the experimental dataset, and the identified evaluation words are quantitatively analyzed.

Keywords: Peer review · Opinion mining · Transfer learning

1 Introduction

China's rapid development has produced numerous scientific research achievements. Because these achievements are the measurement of innovation capability, they have received more and more attention. And scholarly paper is a very important part of the achievements. In order to help scientists do meaningful research instead of pursuing paper numbers, China begins to gradually standardize the scientific research evaluation system. In February 2020, the Ministry of Education and the Ministry of Science and Technology jointly issued "Several Opinions on Regulating the Use of Related Indexes of SCI Papers in Colleges and Universities to Establish Correct Evaluation Guidance". In order to make

Supported by National Natural Science Foundation of China: Project "Research on the Comprehensive Evaluation Model of Papers Based on Citation Polarity and Comment Mining" (61772103); Project "Research on Academic Recommendations Fusion of Multi-source Information" (61976036).

H. Lin et al. (Eds.): CCIR 2021, LNCS 13026, pp. 129–140, 2021.
https://doi.org/10.1007/978-3-030-88189-4_10

scholars get rid of the"SCI-only" influence, the existing evaluation methods of papers are considered, and rational methods should be widely used in current academic evaluation. As an effective and qualitative evaluation of papers, peer review wins scholars' attention.

Peer review is generally regarded as a process in which experts in one or several fields jointly evaluate an activity. The experts adopt the same evaluation standard and the activity belongs to the experts' fields. Peer review is proved to be a very effective evaluation method, so it is often necessary to go through the stage of peer review before submitting to journals or conferences. With the continuous development of open peer review, a large amount of peer review texts begin to appear on the Internet. It provides us an opportunity to explore the characteristics of expert evaluation and better understand peer review. In peer review texts, evaluation information written by reviewers is the most important, and the evaluation words can help us quickly locate the reviewer's evaluation opinions. So it is necessary to extract evaluation words using opinion method. But it is really difficult to go through these texts manually, because it needs not only rich domain knowledge, but also a huge amount of time. Fortunately, artificial intelligence is developing actively. To the best of our knowledge, few people do this task for peer review, so it doesn't have labeled data. Therefore, we choose to add the transfer learning method in order to obtain evaluation words more precisely. When the machine correctly learns the domain knowledge, due to its unique data processing speed, it can quickly and accurately complete the identification of peer review information.

The peer review texts are from ICLR 2018 (International Conference on Learning Representations), which can be found in the OpenReview website. Since there was no annotated dataset for fine-grained peer review texts, as part of this study, our dataset was created by annotating the peer review texts with 47172 words and labeling each word with the type of direct evaluation word, indirect evaluation word, fact word or auxiliary word. We propose a method TransPeerBCC (**Trans**fer Learning-**B**iLSTM-**C**NN-**C**RF for **Peer** Review). We cast the task as a word-level sequence labeling task, utilize BiLSTM-CNN-CRF framework to identify evaluation words, and conduct quantitative analysis on these evaluation words. Through the analysis of evaluation words, we can use direct and indirect evaluation words to distinguish high scores from low score peer review texts, which proves the effectiveness of our method.

To summarize, our contributions include:

- We propose a new task of extracting fine-grained evaluation words. Meanwhile, a useful corpus that facilitates this study is created.
- We use transfer learning to enhance the effect of the experiment, 3 public datasets are used and achieve good results.
- Detailed analysis demonstrates the effectiveness of mining evaluation words and application, which also motivates future research directions.

2   Related Work

2.1   Peer Review

At present, case studies, process improvement studies, and research on new methods of peer review are very common. The new method includes selecting and evaluating the content of journal papers. As a supplement to traditional peer review, it generally includes open review and post-review. However, as a qualitative evaluation method, peer review is less effective than quantitative evaluation [1]. The open peer review mentioned in the previous section provides an opportunity to identify the information in a large amount of peer review texts and conduct quantitative research. The application of sentiment analysis to peer review research has begun to rise, by analyzing the sentiment contained in integral review comments to predict whether a paper will be accepted or not [2]. The result of whether it has been accepted can be used to evaluate the paper more accurately. Hua et al. also make progress in the field of learning peer review texts automatically [3], they define the peer review sentence as different types and present an analysis of them. But as the research deepens, more detailed information are required by scholars, so it is beneficial to investigate the fine-grained information.

2.2   Opinion Mining

Opinion mining is a basic task in the field of natural language processing, which can obtain the required opinions from different types of text [4]. In recent years, the method based on deep learning has been widely used. Ma and Hovy proposed the BiLSTM-CNN-CRF architecture [5]. The model combines character vectors and word vectors to allow the model to cover more input information. After the BiLSTM layer extracts features, the CRF layer normalizes. This is often used in opinion mining. Evaluation word is a kind of opinion, so in this study we can use opinion mining method. Evaluation word recognition for peer review is a new field in which there is less labeled data. In order to make the model fully learn the data features and improve the accuracy of the model, scholars have used the transfer learning method in recent years. In addition to labeling a large amount of data, scholars also propose a method of using transfer learning to transfer the model parameters trained from a large corpus to the target field data [6]. The effective application of transfer learning methods in other emerging fields also lays a solid foundation for the application of peer review evaluation words. Transfer learning is a very popular method at present, and the research of some scholars has good results: The experimental result of Yang et al. shows that public domain datasets can achieve better scores as source data [7]. In order to enhance the model's ability and make the model have better generalization, Zhou et al. proposed to add adversarial discriminator and adversarial training to the transfer learning framework to explore effective feature fusion between high and low resources [8]. Using these methods can make peer review evaluation word recognition achieve better results. In this research, the source data is public domain data, and the target data is peer review text.

3 Task and Data

3.1 Task Description

The task is divided into three parts. The process is shown in Fig. 1.

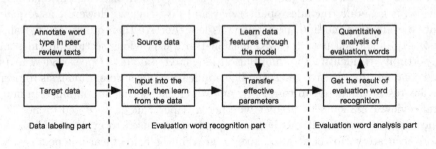

Fig. 1. Flow chart of TransPeerBCC method.

The first part is the data labeling part, which is mainly to process the unlabeled peer review texts into labeled data, which is used as the model's input. The second part is the evaluation word recognition model part, which is mainly to learn the features in the data and complete a trained Model. In the future, the model can identify evaluation words in unlabeled peer review texts. This part is the application part of opinion mining and transfer learning methods. The model method is shown in Fig. 2. The third part is the evaluation words analysis part.

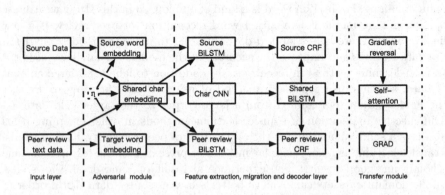

Fig. 2. TransPeerBCC method model framework.

3.2   Dataset

We use the open source peer review dataset that has been labeled with propositional viewpoint categories as an annotation guideline [3]. We first collect all ICLR 2018 peer review data.

Then, based on the analysis of peer review data, the types of evaluation words in peer review are divided into the following four categories: direct evaluation word, indirect evaluation word, fact word and auxiliary word. We use this standard to label all words. Data samples are shown in Table 1. Direct evaluation words can highlight the words used to express the reviewer's direct evaluation of the content of the paper or the writing of the paper. The indirect word can indicate the reviewer's comments on the paper, the modified comment belongs to the reviewer's indirect evaluation of the paper. The words used to show the fact of the paper are called fact word. Auxiliary words refer to the types of words other than the above three words, which are used to help to read and understand.

We maintain a table that maps each ground-truth evaluation words. When we are understanding the meaning of the four word type and checking the reference sentences, We initialize the mapping table by taking the most frequent evaluation words. Then, the evaluation words are assigned into direct evaluation word table and indirect evaluation word table. After that, the words in the table are used to annotate the rest of the peer review text. At the same time, the newly-observed evaluation words are checked, which will be added to the corresponding table only if they have not yet been included. Few examples can be found in the subsequent checking. Most meaningful verbs are annotated manually as fact words, and other words are treated as auxiliary words. Finally, we get 47172 words for the dataset. The direct evaluation table has 418 words, the indirect evaluation table has 97 words.

Table 1. Sample peer review data.

Word type	Example	Sentence for example word
Direct evaluation word	Well-written	The paper is well-written
Indirect evaluation word	Suggest	Hence I suggest you revise your qt set
Fact word	Use	A sequence classifier is also used to tag the presence of topic words
Auxiliary word	All	Are the training and testing sets all disjoint?

4   Evaluation Opinion Mining

4.1   Feature Extraction, Transfer and Decoder Layer

The model architecture used in this part is BiLSTM-CNN-CRF. First, in order to make the input contain more information. CNN (Convolutional neural network) with max pooling is used to process input character features. The output result and the pre-trained word embedding are spliced into the BiLSTM (Bidirectional Long Short-Term Memory Networks). BiLSTM is a type of RNN (Recurrent Neural Network), which is a combination of forward LSTM and backward LSTM. It can save forward and backward text features at the same time, which is suitable for judging word boundary with contextual information. Then CRF model is used here for global normalization, because it contains context information, using the linear chain CRF model as the last layer of the BiLSTM model can obtain the global optimal value more effectively. CRF combines the characteristics of maximum entropy model and hidden Markov model. It is an undirected graph model. For the input sequence x and the label sequence y, define the matching score:

$$s(x,y) = \sum_{i=0}^{l} T(y_i, y_{i+1}) + \sum_{i=1}^{l} U(x_i, y_i) \tag{1}$$

Where l is the length of the sequence, and T and U are learnable parameters. Next, the CRF model is globally normalized:

$$P(y|x) = \frac{e^{s(x,y)}}{\sum_{\widetilde{y} \in y_x} e^{s(x,y)}} \tag{2}$$

Where $\widetilde{y}$ is the predicted label sequence, which maximizes the log probability of the correct label sequence during the training process:

$$log_p(y|x) = s(x,y) - log(\sum_{\widetilde{y} \in y_x} e^{s(x,y)}) \tag{3}$$

This part also contains knowledge sharing mechanism. The knowledge sharing mechanism refers to a method that shares the knowledge learned by the model in the source data to the model for target data. According to the experience of previous researchers, if the labels are similar, all parameters of the model will be shared. Otherwise, only a few parameters will be shared. When selecting experimental subjects, there is currently no appropriate public data of evaluation words as the source data, so this study chooses to share part of parameters.

4.2   Transfer Module

The transfer Module contains three parts: gradient reversal module, self-attention mechanism and generalized resource adversarial discriminator (GRAD). These three parts have been proved to make the model produce better transfer effects. The functions of these three parts are described below.

The gradient reversal module is mainly used between the feature extractor and the domain classifier. In the backward propagation process, the gradient of the domain classification loss of the domain classifier is automatically inverted before being back propagated to the parameters of the feature extractor [9].

The self-attention mechanism is a kind of attention mechanism. The attention mechanism is an algorithm that calculates the importance of different parts of a certain thing. It allocates more attention to the key parts of things [10].

GRAD can help the model identify the source data, make the feature representations from the source field and the target field more compatible. It uses different adaptive weight coefficients to maintain the balance of source data and target data and control the contribution of a single sample to the loss function. In order to calculate the loss function of GRAD, the output sequence of the shared BiLSTM is first encoded as a single vector by the self-attention module, and then projected to a scalar by linear transformation r:

$$\ell_{GRAD} = -\sum_i \{I_{i\in D_s}\alpha(1-r_i)^\gamma log r_i + I_{i\in D_T}(1-\alpha)r_i^\gamma log(1-r_i)\} \tag{4}$$

Among them, $I_{i\in D_s}$, $I_{i\in D_T}$ are the identity functions representing the source of the sequence, α is the weight coefficient that keeps the source data and the target data balanced. r_i^γ (or $(1-r_i)^\gamma$) controls the contribution of a single sample to the loss function.

4.3   Adversarial Module

Before the model learns the parameters, adversarial samples are added to the embedding layer to allow the model to learn with subtle disturbances. Adversarial samples x_{adv} are constructed by the formula:

$$x_{adv} = x + \eta_x \tag{5}$$

η_x is calculated by adding a norm term ϵ to the loss function:

$$\eta_x = \epsilon \frac{\nabla \ell(\Theta; x)}{\| \nabla \ell(\Theta; x) \|_2} \tag{6}$$

5   Experiment

5.1   Peer Review Mining

The source data is selected from the public domain English named entity recognition dataset with good transfer level: Ontonote5.0 dataset [11], CoNLL2003 dataset [12], and WNUT2016 dataset [13]. The target data is divided into direct evaluation word, indirect evaluation word, fact word and auxiliary word. These words use the BIO sequence labeling method (Evaluation words include single evaluation word and evaluation phrases, such as make sense, not clear.) B represents the beginning of the evaluation word, and I represents the word within the evaluation word, words marked as O are auxiliary words for comprehension. The model uses ELMo word vector [14], and uses accuracy, precision, recall and F1 value as the evaluation measurements for this experiment. The experimental results are shown in Table 2.

Table 2. Data transfer experiment results from different sources.

Source data	Accuracy	Precision	Recall	F1
OntoNotes 5.0	97.40	78.21	66.30	71.76
CoNLL2003	96.89	68.97	65.22	67.04
WNUT2016	97.35	76.19	69.57	72.73
–	96.69	63.83	65.22	64.52

It can be seen from the results that transfer learning is useful for our task. The experiments with transfer learning method are improved significantly. Different source data affects in varying degrees. WNUT2016 achieves the best F1 scores, maybe because it has closer expressions with peer review texts.

5.2 Quantitative Analysis of Evaluation Words

First, we extracted the stems of evaluation words. Then we counted the word frequency and made a co-occurrence matrix. Finally, we imported the co-occurrence matrix into Gephi software, deleted nodes with a degree less than 50, and obtained the direct evaluation word co-occurrence network, as shown in Fig. 3. It can be seen from the figure that most of the evaluation words are adjectives for evaluating the study design and experimental results. And the 5 most common direct evaluation words, their frequency and examples are shown in Table 3.

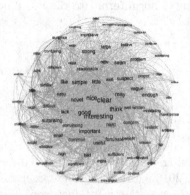

Fig. 3. Direct evaluation word co-occurrence network.

Table 3. Common direct evaluation words, frequency and examples.

Direct evaluation word	Frequency	Examples
Clear	96	1. The explanation is clear 2. This would allow the author to clearly delineate their contribution 3. It is unclear how crucially this additional assumption is required in practice
Interesting	63	1. It is perhaps interesting that one can make deep learning learn to cooperate 2. The authors draw some interesting conclusions 3. All in the paper is interesting
Good	42	1. The experiment is good 2. The area under this curve allows much better to compare the various methods 3. The quality of the paper is good, and clarity is mostly good
Think	31	1. I think this paper has many issues 2. I don't think that these pictures are the most commonly known example 3. I think you mean the song 's section
Hard	21	1. The paper is hard to read 2. It is hard to understand the experiment 3. As out of distribution samples are hard to obtain

In summary, peer review texts contain a plenty of direct evaluation words. According to the habits and the interest level of reviewers to the paper, peer review texts with different styles can be obtained. Reviewers will say that one paper is interesting and novel, but they will not directly say that another paper is not interesting and not novel. Reviewers will use common and simple to express their thought.

In peer review texts, reviewers often evaluate both paper writing and paper content. The words used for paper writing are often well-written, hard to read. As for the paper content, reviewers often suggest that the paper is too easy, the idea is very interesting, and the study is clear and so on. At the same time, in the test dataset, the model effectively recognized the words that never appeared in the training sample. The words represented by the direct evaluation words are taken as examples, as shown in Table 4.

Table 4. The recognized words and example sentences.

Recognized word	Example
Heuristic	The proposed techniques are heuristic
Interpretable	The sequence of subtasks generated by the policy are interpretable

Indirect evaluation words are often modal verbs, which are used to give revised opinion to authors. The most common indirect evaluation words are shown in Table 5.

5.3 Application

The TransPeerBCC method proposed in this paper can effectively identify the direct and indirect evaluation words in peer review texts. Through the direct evaluation words, the reviewer's evaluation summary can be located quickly, which helps to predict the reviewer's scores for the paper better and infer the

Table 5. Common indirect evaluation words, frequency and examples.

Indirect word	Frequency	Example
Should	92	1. An attempt to explain what Word2Vec is doing should be made with careful experiments over many relations and hundreds of examples
		2. Other hyperparameters should also be considered
		3. There should be some ablation study here
Would	53	1. This work would benefit from careful copy editing
		2. I would expect more explanations
		3. A better comparison would be to plot the performance of the predictor of S against the performance of Y for varying lambdas
Could	45	1. Could you provide how much extra computation is needed for that model?
		2. They could verify how much accuracy improvement is due to the insensitivity to order in filtering expressions
		3. I think this section and the previous could be combined

final acceptance/rejection comments. Through the indirect evaluation words, the author can clearly know the reviewer's revised views, discover the shortcomings of his paper in time and make revisions and improvements. This study selects two common peer review text segments as shown in Table 6. {} identifies direct evaluation words, and [] identifies indirect evaluation words. It can be seen from Table 6 that by quickly locating the evaluation words, compared with the right, the left peer review contains more positive evaluation, and the deficiencies considered by the reviewers are also some minor flaws.

Table 6. Examples of evaluation word recognition applications

Review1 Score:7	Review2 Score:3
In this paper, the author... I {enjoyed} reading this paper. It is a very {interesting} set up, and a {novel} idea. A few comments: The paper is {easy to read}, and largely {written well}. The article is {missing} from the nouns quite often though. So this is something that [should] be amended... There are a few spelling {slip ups}...	This work... It's {not clear} what can be said with respect to the convergence properties of this class of models, and this is not discussed... Unfortunately the only quantitative measurements reporter are Inception scores, which is known to be a {poor} measure; Frechet Inception distance or log likelihood estimates via AIS on some dataset [would] be more convincing...

6   Conclusion

This paper proposes an evaluation word recognition method TransPeerBCC for peer review texts. It adopts the opinion mining method based on transfer learning to discover the effective information in the peer review texts. Through this method, the evaluation words in the peer review texts can be effectively identified. Through the evaluation words, people can quickly understand the peer review and better know the opinions of the reviewers. The proposed model can help humans understand numerous peer reviews in a short period of time. At the same time, this method is currently only a preliminary recognition of the evaluation words. Future research can have a deeper understanding of the insights put forward by reviewers.

References

1. Yang, Y., Zha, Y.: Enlightenment and reference from the experience of foreign scientific and technological talents: a study based on Britain, the United States and Germany. Sci. Manag. Res. **38**(1), 160–165 (2020)

2. Wang, K., Wan, X.: Sentiment analysis of peer review texts for scholarly papers. In: The 41st International ACM SIGIR Conference on Research and Development in Information Retrieval, pp. 175–184 (2018)

3. Hua, X., Nikolov, M., Badugu, N., Wang, L.: Argument mining for understanding peer reviews. arXiv preprint arXiv:1903.10104 (2019)

4. Lippi, M., Torroni, P.: Argument mining: a machine learning perspective. In: Black, E., Modgil, S., Oren, N. (eds.) TAFA 2015. LNCS (LNAI), vol. 9524, pp. 163–176. Springer, Cham (2015). https://doi.org/10.1007/978-3-319-28460-6_10

5. Ma, X., Hovy, E.: End-to-end sequence labeling via bi-directional LSTM-CNNs-CRF. arXiv preprint arXiv:1603.01354 (2016)

6. Li, M., Li, Y., Lin, M.: Review of transfer learning for named entity recognition. J. Front. Comput. Sci. Technol. 15(2), 206–218 (2020)

7. Yang, Z., Salakhutdinov, R., Cohen, W.W.: Transfer learning for sequence tagging with hierarchical recurrent networks. arXiv preprint arXiv:1703.06345 (2017)

8. Zhou, J.T., et al.: Dual adversarial neural transfer for low-resource named entity recognition. In Proceedings of the 57th Annual Meeting of the Association for Computational Linguistics, pp. 3461–3471 (2019)

9. Ganin, Y., Lempitsky, V.: Unsupervised domain adaptation by backpropagation. In: International Conference on Machine Learning, pp. 1180–1189. PMLR (2015)

10. Vaswani, A., et al.: Attention is all you need. In: Advances in Neural Information Processing Systems, pp. 5998–6008 (2017)

11. Hovy, E., Marcus, M., Palmer, M., Ramshaw, L., Weischedel, R.: OntoNotes: the 90% solution. In: Proceedings of the Human Language Technology Conference of the NAACL, Companion Volume: Short Papers, pp. 57–60 (2006)

12. Sang, E.F., De Meulder, F.: Introduction to the CoNLL-2003 shared task: language-independent named entity recognition. arXiv preprint cs/0306050 (2003)

13. Mishra, S., Diesner, J.: Semi-supervised named entity recognition in noisy-text. In: Proceedings of the 2nd Workshop on Noisy User-Generated Text (WNUT), pp. 203–212 (2016)

14. Peters, M.E.: Deep contextualized word representations. arXiv preprint arXiv:1802.05365 (2018)

Evaluation of Learning Effect Based on Online Data

Zhaohui Liu[1]([✉]) [iD], Hongfei Yan[1,3]([✉]), Chong Chen[2] [iD], and Qi Su[1] [iD]

[1] Peking University, Beijing, People's Republic of China
{lzh0290,yanhf,sukia}@pku.edu.cn
[2] Beijing Normal University, Beijing, People's Republic of China
chenchong@bnu.edu.cn
[3] National Engineering Laboratory for Big Data Analysis and Application
Technology, Center for Big Data Research, Peking University,
Beijing, People's Republic of China

Abstract. With the development of Internet technology, online learning has gradually been widely used, which has further enriched the teaching methods. However, the indicators provided by online learning platforms to characterize students' learning behaviors are still relatively small, it is hard to effectively evaluate students' learning effects and provide personalized education for students. While students produce a large number of online data records during the learning process, if these data can be used to evaluate the learning effect of students and find indicators that are significantly related to the learning effect, it can help teachers know the learning status of students and provide them with more targeted guidance. The framework of this article includes: firstly collect online data of students' learning, and determine major types of factors that affect the learning effect. Next, the original data is processed, and new features are constructed from the original data according to the factors that affect the learning effect. Then use several different methods to select the features that have the most significant impact on the learning effect, and use several regression methods to predict students' academic performance. Finally, this article selects important features that are more relevant to learning effects, studies the relationship between these features.

Keywords: Data mining · Learning effects · Online data

1 Introduction

Many schools try to combine online and offline teaching methods to form SPOC courses (small private online courses). This course is only open to a certain group of students, and offline teaching will also be carried out in addition to online courses. This type of teaching not only retains the flexibility of online teaching, but also improves student interaction and participation compared with pure

Supported by NSFC Grant 61772044; MSTC Grant 2019YFC1521203; Peking University Grant 2020YBC01.

H. Lin et al. (Eds.): CCIR 2021, LNCS 13026, pp. 141–154, 2021.
https://doi.org/10.1007/978-3-030-88189-4_11

online teaching, improves the degree of personalized teaching. However, there are still some problems to be solved for this teaching form. First of all, there are numerous online behaviors and interactive data. It's very useful to evaluate students' learning situations efficiently with these online data, and it's also the key prerequisite for more accurate positioning of each student and corresponding guidance. Secondly, for the many characteristics of students in the learning process, which factors will have a more obvious connection or influence with the students' learning effect. Exploring these important characteristics can, on the one hand, evaluate students more accurately, on the other hand, it can also be used to improve teaching methods.

1.1 Related Work

At present, scholars use different methods to analyze students' online learning behaviors and the main types of research work include:

1) Analyze the characteristics of single dimension and put forward opinions. Zong Yang and others collected student behavior data on the MOOC platform, and selected 18 indicators such as the number of course pages viewed and then use Logistic regression to predict whether the students' grades are qualified [1]. Sharma and others used sensors to collect students' blood pressure, heart rate, and other body data during online learning, and used Hidden Markov Models to predict whether students were engaged during class [2].

2) Propose corresponding analysis methods from the perspective of the application of the learning platform. For example, Hua Yanfeng et al. proposed a learning analysis application model composed of elements such as learning process, learning environment [3]. The "learning dashboard" proposed by Zhang Zhenhong, as a learning support tool, can be used to obtain feedback information in teaching [4].

3) According to the behavior characteristics of students, analyze and predict their learning outcomes. For example, by issuing questionnaires, Cao Mei counted the frequency of 25 behaviors of students, and analyzed the correlation between different behaviors and learning effects through linear regression [5]. Jiang Zhuoxuan and others selected two basic behaviors: watching videos and submitting quizzes, using thresholds to divide learners into different types, analyzing time-related behavior patterns. Finally, the six indicators in courses are selected, and predict whether the learner will pass the course [6].

1.2 Factors Affecting the Quality of Online Learning

At present, many researchers have explored the factors that affect the learning effect, which mainly include the students' own factors and various external factors [7]. Only self-factors related to learning effects are considered for research, and self-factors that are easy to portray through behavioral data can be divided into:

(1) Devotion: It describes the student's effort to study on the learning platform. Devotion is directly related to factors such as the number of times students study on the platform, the time spent, the completion of homework, etc.

(2) Learning habits: It is mainly reflected in the fact that students tend to study at the time they are accustomed to, and there is a certain regularity in the time of submitting homework, watching videos, etc. Regular learning can also enhance students' sense of experience and improve the quality of learning [8].

(3) Learning efficiency: Different students have differences in knowledge reserves, learning abilities and attitudes, so the time required for understanding and mastering the knowledge, completing various tasks is also different.

(4) Interaction situation: In the process of online learning, communication is one of the important ways to solve problems and puzzles. In addition, students with higher learning levels are also more capable of answering and solving other students' questions, and may be in a more important position in social networks.

(5) Subjective emotions and feelings: Students have different emotions and feelings in the learning process, and they also have subjective evaluations of their learning situation. Different feelings may affect the learning effect.

1.3 Contributions

This article use methods in data mining, machine learning, first collects various data on students' online learning, and then combines the factors that may affect the learning effect in related research, constructs a large number of features from the original data, and uses them to characterize students' learning behaviors and characteristics. Furthermore, the students' course performance is used as an index to evaluate the learning effect, a model is established to explore the relationship between each feature and the learning effect, and a model and index that can evaluate the learning effect of the student are obtained. Finally, this article selects these characteristics that are closely related to the learning effect, studies the relationship between these characteristics.

The achievements are as follows: (1) Rich behavioral characteristics are constructed from the original data, which can more clearly understand and describe the characteristics of students' online learning behaviors. (2) According to the experiment, some features that are closely related to the learning effect are selected, and corresponding models are constructed. Using these features and models can assist in evaluating the learning status of students and help to realize personalized teaching. (3) Explore the impact of some important characteristics, verify some people's concepts and cognitions on learning based on data.

2   Online Learning Data Processing

2.1   Data Sources

This article finally choose "Introduction to Computing" as the research object, and choose the Class12 and Class13, which two classes set up in the Canvas learning management system. Among them, Class12 is an offline class, students are also studying online and completing assignments. A total of 126 people finally completed the course and obtained grades; Class13 is a fully online teaching class, all teaching is carried out online, and 36 students complete the course. The teaching content of the two classes is the same, and the specific online data sources can be divided into:

(1) Canvas teaching platform: Online teaching relies on the learning management system (LMS). All student behaviors on the platform will be recorded and stored in the system's database. A total of 139 tables in the system database store various types of information.
(2) Questionnaire: During the important time such as the beginning, mid-term, and end of the semester, four questionnaire surveys were conducted among students, including learning experience, programming knowledge and mastering of algorithm, etc. At the same time, a questionnaire survey was conducted with the students most familiar with.
(3) Course WeChat group: The two classes have their own course groups. Through the WeChat message, rcontact, and chatroom tables, all the chat records of the course group are obtained in csv format.

2.2   Data Processing of Canvas Teaching Platform

In the database of the Canvas teaching platform, a series of table join operations are performed through SQL statements to obtain the original behavior data of the students on the teaching platform, the record contains information such as the behavior time, the requested url, user id, the category corresponding to the request, and session_id, and a total of 214,442 behavior records. According to the factors that affect the learning effect, use the original data to construct new attributes, some of the obtained attributes are shown in Table 1:
The definition and acquisition methods of some attributes are as follows:

(1) Number of logins: Use the re-divided session segment number as the number of student logins to the platform, which can be obtained by counting the number of session_id fields in the behavior record.
(2) Online time: The duration of each login is the duration of the session. The sum of all login durations of a student is the online time.
(3) The number of times of watching the live broadcast: Judging according to the content of the url field in the behavior record, the record containing the live broadcast address in the url is regarded as a live watch behavior, and the behaviors such as watching video, browsing ppt, etc. are similarly judged according to the url field.

Table 1. Different behaviors and corresponding attributes

Behavior	Corresponding attributes
Login to the platform	Number of logins, average interval, online time, etc.
Watch course videos	Number of views, average interval, fluctuations in interval, etc.
Browse courseware	Number of views, average interval, interval fluctuations, etc.
Submit homework	Number of submissions and comments, assignment score, etc.

(4) The average interval of watching live broadcast: the average number of days between watching the live broadcast of the course, and calculate the average of all the interval days by counting the date when the live watching behavior occurs. The same can be calculated for other types of behaviors.

(5) The number of days to submit homework in advance: the length of time between the due date of the homework and the date when the student actually submitted the homework, calculated from the homework submission time and due time obtained in the database.

2.3 Data Processing of the Questionnaire

During the course of the course, four questionnaire surveys of students in each class were conducted, as well as a survey of the most familiar classmates. The missing values in the questionnaire are filled in first [9]. In this paper, four filling methods are selected for experiment and the results are compared [10]. First, select all the questionnaire records without missing values from the original data as the experimental data set, randomly generate missing values on the experimental data set according to the missing ratio of the original data. The four methods used are: (1) Mode filling (2) K-nearest neighbor filling (3) Random forest filling method and (4) Multiple imputation method.

The results on the experimental data set are as follows. The k nearest neighbor is $k = 5$, and the number of decision trees in the random forest is $n = 800$. It can be seen from the filling results that the filling effects of the four methods are not much different. In summary, the k-nearest neighbor filling method is used to process the missing values in the questionnaire, and 114 questionnaire questions in Class12 are obtained (Table 2).

Then, according to the questionnaire of "the classmate you are most familiar with", construct the social network in the class and study the relationship between the attributes of students in the social network and the learning effect [11]. This article regards students as the nodes of the social network, and the familiar relationship between students as the connection of the nodes. If student A thinks that student B is one of the most familiar students, then construct A connect to B. After the construction is completed, the final calculated

Table 2. The filling effect of different methods on the experimental data set

Filling method	Error rate	MSE
Mode filling	0.223	0.3369
K-nearest neighbor filling	0.214	0.3069
Random forest filling	0.222	0.3123
Multiple imputation method	0.253	0.3708

attributes are: (1) Out degree: refers to the total number of classmates that the student thinks is familiar, (2) In degree: the total number of times the student is considered familiar by other students. (3) Between Centrality: measure the degree of a node as a bridge and the ability to communicate with different groups. (4) Closeness Centrality: measure the overall status of students in the social network.

2.4   Data Processing of Course WeChat Group

Class 12 students have a total of 7276 text chat records, while only 484 in Class 13, which is far less active than Class 12. Since there are few chat records in Class 13, no other features are counted except the number of speeches. Some of the constructed attributes are shown in Table 3:

Table 3. Some features of WeChat records

Attributes	Meaning
Number of speeches	Number of students' speeches in the course group
Number of words spoken	Number of words spoken by students
Number of highly engaged speeches	Number of speeches that are highly relevant to the course content
Number of Thoughts	Number of speeches reflecting students' thinking

Some attributes are obtained in the following ways: (1) Number of confusions: If the speech contains words such as "why", "asking", "how to do" and other expressions of doubt and confusion, the number of confusions is increased by one. (2) Number of expressions of emotions: If the speech contains the expressions that come with WeChat, corresponding to words such as "face cover", "fist", "smirk", the number of expressions of emotions is increased by one.

Then use a pre-constructed dictionary to calculate the corresponding attributes according to whether the interactive record contains the terms in the dictionary. The construction of each attribute dictionary is as follows: first, according to the speech records after word segmentation, count the frequency of all spoken words, and remove the stop words. In the 522 words that appear more frequently, select words related to attributes and add them to the dictionary.

3   Evaluation of Learning Effect

A total of 272 features were obtained from Class 12, 114 of which were questionnaire answers, and 158 behavior features were from Canvas platform and WeChat. The original features in the questionnaire belong to students' subjective self-perception, while the behavior features constructed from other data belong to the objective description of students. Therefore, first use two types of features to predict the learning effect, and compare the impact of two types of features, and then select the important features of the two types for evaluation.

This article uses the student's course score as an indicator of the actual learning effect, and uses four different regression prediction methods to use each feature as the variable to predict the learning effect and compare it with the indicator of the score. Then uses R-Squared to measure the effect of prediction, and cross-validation is used to prevent overfitting [12].

3.1   Linear Regression

First, select features base on Adjusted R-Squared. $AdjustedR - Squared = 1 - -\frac{(1-R)(n-1)}{(n-p-1)}$. R is the R-Squared value, n is the number of samples, and p is the number of features. Adjusted R-Squared can punish the added non-significant variables. Only when the added variable has a more significant effect, the Adjusted R-Squared value will increase [13]. Take advantage of this nature, first, sort all the features according to the absolute value of the correlation coefficient from largest to smallest. Then select a new feature in order for ols regression, repeatedly loop until it is satisfied that each new feature increases Adjusted R-Squared, and the final selected feature set is obtained.

Using the above method to select behavior characteristics, a total of 28 characteristics are finally obtained. Adjusted R-Squared is 0.591, When performing linear regression, the average R-squared value of the 10-fold cross-validation is 0.318, and the R-squared value of the entire data set is 0.538. The results of the regression of the selected features are shown in Table 4.

The relationship between some of these characteristics and grades is the same as usual perceptions. For example, students who make more highly engaged speeches on WeChat are more close to the centrality, they participate in the interaction more actively; students who review videos more often, watch the live broadcast at a smaller average interval are more engaged in learning; students who take less time to complete their homework and check the homework as soon as possible, their learning efficiency is relatively higher. Such students are more likely to achieve better results. There are also some characteristics, such as the total number of behaviors on Monday, at 11 o'clock and 19 o'clock, etc., which are related to the study habits of students, indicating that study habits will

Table 4. Regression results of behavior features

Features	Regression coefficients
Closeness centrality	9.2418
Number of days to browse the document	−19.4915
Average interval of watching videos	5.4964
Total number of watching videos	4.2361
The total number of online behaviors for more than one hour in morning	5.9280
In-degree	−6.3398
Out-degree	−10.6005
Increase in total number of behaviors in the fifth week compared to the fourth week	3.2217
Total number of behaviors at 11 am	−4.1506
Total number of behaviors at 19 pm	−4.8027
Total number of behaviors in the fifth week	4.6802
Total number of behaviors on Monday	−5.4969
Average number of behaviors per login	−8.5713
Maximum number of days between logins	0.9258
Coefficient of variation of the interval between watching live	1.4182
Number of homework comments	2.0505
Number of WeChat highly participate speech	8.1698
Total number of viewed documents	−0.1785
Number of times it takes less than the average of classmates to complete the homework	11.4611
Number of documents browsed	3.2175
Average time from posting assignments to the first viewing of assignments	−14.3375
Average interval of watching live broadcasts	−9.2876
Number of times first browse the homework is greater than average	−8.1989
Average number of words spoken on WeChat	4.1729
Number of online behaviors for more than one hour during weekends	−5.8459
Average time to complete homework	−0.2962

also have an impact on academic performance. Some of the characteristics can be inferred and reasonable explanations can be found. For example, the total number of behaviors and the growth rate in the fifth week should be related to the curriculum. Since the fifth week starts to explain the more difficult content such as algorithms, students may have a better grasp of the difficult points if more investment is made in this period, so the final score will be higher. However, there are also some characteristics that are difficult to explain. For example, the degree of out, degree in, and the average number of behaviors per login are negatively correlated with performance, there may be interactions with other characteristics. In general, the model has good explanatory properties.

Then, using Adjusted R-Squared for feature selection on the questionnaire features, a total of 26 features are obtained, and Adjusted R-Squared is 0.723. When performing linear regression, the average R-squared value of 10-fold cross-validation is 0.520, and the R-squared value of the entire data set is 0.721. The result is shown in Table 5. The numbers in parentheses represent the options of question. 0 and 1 represent no and yes respectively, while 1, 2 represent yes and no, and 1–5 represent multi-valued questions of ordered categories.

It can be seen from the regression results that students believe that whether they have mastered the knowledge points of the course are closely related to their final grades. For example, students who think they have mastered matrix operations, dynamic programming, depth-first search and greedy algorithms may have better final grades. Among them, knowledge points such as greedy algorithms, matrix operations and basic Python grammar have a more obvious impact on the results, and may account for a larger proportion of the course exam. Students who recognize their own learning input and believe that they have learned a lot from the course are also more likely to achieve better results. At the same time, the effect of using the characteristics of the questionnaire alone to predict student performance is significantly better than the effect of using behavioral characteristics alone. It is because the characteristics of the questionnaire mainly reflect the subjective feelings and self-evaluation of students, indicating that students' self-evaluation has high accuracy and good effect in evaluating learning effects. And since the students of Class12 learn with a combination of online and offline methods, online learning data can not represent all learning behaviors, so the characteristics of the questionnaire have a better results.

There are 54 selected features in the above two categories. For further selection, 35 features were obtained, of which 17 were extracted from behavioral data, and 18 were questionnaire features. Adjusted R-Squared is 0.822, when performing Linear Regression, the average R-squared value of the 10-fold cross-validation is 0.677, and the R-squared value of the entire data set is 0.854. It can be seen that the effect of using the two types of features to predict together has been significantly improved, and compared with the case of using one type of feature alone, the number of features used has not increased significantly, indicating that the model has achieved better prediction results. The regression results are not shown for the limit of space.

From the results, it can be seen that there is a close relationship between whether to master certain knowledge points and the learning effect. It is believed that students who have not mastered the knowledge points may achieve lower grades. This shows that the students' assessment of their own mastery is more consistent with the actual situation, which is very helpful for assessing and under-

Table 5. Regression results of questionnaire features

Features	Regression coefficients
Use Mac system (0,1)	1.8140
Attend class in the classroom in October (0,1)	2.5728
I wanted to have this instructor (0,1)	−1.8907
Know the memory structure diagram when the program is running (1,2)	1.2808
Materials posted in Canvas and WeChat groups are highly targeted (1,5)	−1.2361
Master the use of OpenJudge test data to test your programs (1,2)	1.5070
Learn a lot of new knowledge in the October course (1–5)	1.0127
Mastered the basic grammar of Python at the end of the term(1,2)	−7.7202
Time investment of a semester is worthwhile (1–5)	2.3059
Have used leetcode (0,1)	4.8173
Practice typing and English (0,1)	1.1929
Attend class in the classroom in December (0,1)	−2.8811
Master one-dimensional array copying (1,2)	1.2450
The optional assignments in November are challenging, can be completed with hard work (1–5)	−1.0435
Master the dual pointer algorithm at the end of the term (1,2)	−0.9761
Master matrix operations at the end of term (1,2)	−6.8351
Master the idea of dynamic programming algorithm in November (1,2)	−1.8878
Need to explain the code of optional assignments in November (1,2)	2.7201
At the end of term, the total number of passed questions on other programming platforms	−0.6924
Master the algorithm ideas of dfs/bfs at the end of the term (1,2)	−1.5659
Master the problem of OJ 1174 in November (1,2)	−4.4216
Participate in the final exam of Class12 (1,2)	−5.1233
Need to explain the code of must-do assignments in November (1,2)	2.4161
Master the Greedy Algorithm at the end of the term (1,2)	−10.2680

standing the students' learning situation. In addition, the learning habits of certain periods also have an impact on the learning effect, especially in the difficult teaching period of the course. Students with greater input may have a better grasp and eventually achieve better results. The students who check the assign-ment as soon as possible after the assignment is released are more motivated to learn and the learning effect is better.

3.2 Lasso Regression

Least absolute shrinkage and selection operator (Lasso) Regression construct a penalty function to compress the coefficients of some variables and make some regression coefficients become 0, so as to achieve the purpose of variable selection and enhance the generalization ability of the model. For high-dimensional feature data, Lasso regression can reduce the dimensionality and prevent overfitting [14].

As for behavior features, the maximum average R-squared value of the 10-fold cross-validation is 0.430, and the R-squared value on all data sets is 0.731, and a total of 30 features are selected. As for questionnaire features, the average R-squared value of the 10-fold cross-validation was 0.489, and the R-squared value on all data sets was 0.803, a total of 32 features were selected.

For reasons of space, the results of regression prediction is not listed. Com-pared with the results of linear regression, it is found that the two have selected a large part of the same characteristics. For example, when selecting behavioral characteristics, the two methods selected 16 identical characteristics, such as closeness centrality, average number of words spoken on WeChat, average inter-val for watching live broadcasts, and the number of browsed documents. There are also correlations and similarities in other features, such as the total number of words in the homework comments selected by Lasso regression, the number of homework comments selected by linear regression. When selecting the char-acteristics of the questionnaire, both methods choose whether to master certain knowledge points, but the selection of knowledge points is not same.

Lasso regression was performed again on a total of 62 features of the two categories. The maximum average R-squared value of the 10-fold cross-validation was 0.629, and the R-squared value on all data sets was 0.860. A total of 39 features were selected. Among them, there are 9 behavioral characteristics and 30 questionnaire characteristics. The proportion of questionnaire characteristics is significantly larger, which shows that the characteristics of the questionnaire have a significant impact on the learning effect.

3.3 Random Forest Regression

The process of feature selection using the random forest algorithm is: first per-form regression prediction on all features, obtain the average R-squared value of cross-validation, and the importance weight of each feature in the prediction, remove the feature with the smallest weight, and make the remaining features as the feature set to be filtered. Random forest regression is used again for the feature set to be screened, and the new weight corresponding to each feature and the average R-squared value of cross-validation are obtained, and the loop continues until all the features are screened. Finally, the feature set correspond-ing to the maximum average R-squared value of the cross-validation is taken as the selected feature. Finally, 12 features were selected from all the features, the average R-squared of cross-validation was 0.593, and the R-squared on the entire data set was 0.943.

Further, the weights obtained by the random forest algorithm are used to perform feature selection [15] for linear regression. Using this method to select a total of 35 features, the average R-squared value of the 10-fold cross-validation is 0.718, and the R-squared value on all data is 0.819. Observing the selected features, it is found that the features selected by this method and the previous three options have a large part of the intersection, only the total number of behaviors at 9 am, and the other 6 items do not appear in the previously selected feature set. Among the selected features, the number of behaviors in different time periods reached 7 features, which made the model's explanatory performance decrease (Table reftab6).

Table 6. Regression effect of different methods

Method	Number of features	Average R-squared of cross-validation	R-squared on all data
Linear regression with adj. R-squared	35	0.677	0.854
Linear regression with random forest	35	0.718	0.819
Lasso regression	39	0.629	0.860
Random forest regression	12	0.593	0.943

3.4 Result Analysis

As for Class 13, Lasso regression was used for prediction. A total of 25 features were selected. The average R-squared value of cross-validation was 0.727, and the predicted R-squared value on all the data was 0.998. Among them, 16 items are characteristics of the questionnaire. The proportion of behavioral characteristics has increased compared with the prediction of the Class12, it is because Class13 adopts a completely online learning method, the online behavior data more accurately describes the learning behavior of students. The characteristics of the questionnaires in the two classes both account for a large proportion, indicating that no matter which teaching method is adopted, the self-evaluation and feedback of students are important evaluation indicators.

Comparing the prediction effects of various methods, the random forest regression selects the least number of features, and it is the closest to the true value on all data, but the effect of cross-validation is worse than other methods. At the same time, random forest regression is nonlinear regression, the exact relationship between features and results cannot be obtained. The other three methods have better cross-validation effect and generalization ability, and get the quantitative relationship between characteristic variables and performance, which is more concise, and most of the features in these three methods are interpretable. Overall, they have achieved good results in predicting the learning effect of class 12 students.

Comparing the selected characteristics of the two classes, it is found that some of the characteristics are common influencing factors. For example, the number of videos watched and the status of mastering important knowledge are positively correlated with the results, while the time to complete the homework are negatively correlated with grades. Compare features selected using different methods, and consider feature variables selected in multiple models as important features. These features are also the characteristic variables finally selected to characterize students' learning behavior, including the average time to complete homework, the average time from homework assignment to the first browsing, the number of days to browse the document, the number of times to watch the live broadcast, and self-assessment of important knowledge points, etc.

4 Conclusion

The average time to complete the homework, the average time to browse the homework for the first time, the number of interactive speeches, and the students' self-evaluation of the mastery, etc. are selected through experiments, which can make corresponding assessments of the learning situation and help teachers to achieve personalized teaching. In the future, this process can be promoted in a wider range to test the effects of the methods and the selected indicators.

In addition, the quantitative calculation of the relationship between certain factors and the learning effect based on the data supports the qualitative views and opinions that some people have formed, such as the relationship between the degree of active participation in discussions, the time spent completing homework, etc. and the learning effect. At the same time, it is also found that the questionnaire is effective in obtaining the learning status of students, and there is a close connection between the behavioral data of students and the results of the questionnaire. The two can be mutually confirmed, and the combined use helps to better evaluate the learning status of students.

References

1. Zong, Y.: A logistic regression analysis of learning behaviors and learning outcomes in MOOCs. Distance Educ. China **36**(004), 73–74 (2016)
2. Sharma, K., Papamitsiou, Z., Olsen, J.K., et al.: Predicting learners' effortful behaviour in adaptive assessment using multimodal data. In: 10th International Learning Analytics and Knowledge (LAK) (2020)
3. Hua, Y.: Construction of a multivariate concentric learning analysis model based on MOOCs. J. Dist. Educ. (005), 104–112 (2014)
4. Zhang, Z.: Learning dashboard: a new learning support tool in the time of big data. Modern Distance Educ. Res. **3**, 100–107 (2014)

5. Cao, M.: College students' blended learning behavior and its mechanism. Modern Distance Educ. **1** (2020)
6. Jiang, Z., Zhang, Y., Li, X.: Learning behavior analysis and prediction based on MOOC data. J. Comput. Res. Dev. **3**, 614–628 (2015)
7. Li, L.: Analysis of psychological factors affecting the quality of online learning. Adv. Mater. Res. **926–930**, 4461–4464 (2014)
8. Andresen, M.A.: Asynchronous discussion forums: success factors, outcomes, assessments, and limitations. J. Educ. Technol. Soci. **12**, 249–257 (2009)
9. Cismondi, F., Fialho, A.S., Vieira, S.M., et al.: Missing data in medical databases: impute, delete or classify? Artif. Intell. Med. **58**(1), 63–72 (2013)
10. Li, L., Yang, H., et al.: Method of processing missing values based on clinical data sets. China Digit. Med. **13**(04), 13–15+85 (2018)
11. Eckles, J.E., Stradley, E.G.: A social network analysis of student retention using archival data. Soc. Psychol. Educ. **15**(2), 165–180 (2012)
12. Ricci, L.: Adjusted R-squared type 'measure for exponential dispersion models. Stat. Probab. Lett. **79**(17–18), 1365–1368 (2010)
13. Zhang, Y., Yang, Y.: Cross-validation for selecting a model selection procedure. J. Econometr. **187**(1), 95–112 (2015)
14. Sinha, P., Verma, A., Shah, P., et al.: Prediction for the 2020 United States presidential election using machine learning algorithm: Lasso regression. MPRA Paper (2020)
15. Uddin, M.T., Uddiny, M.A.: A guided random forest based feature selection approach for activity recognition. In: International Conference on Electrical Engineering and Information Communication Technology. IEEE (2015)

Self-training vs Pre-trained Embeddings for Automatic Essay Scoring

Xianbing Zhou[1], Liang Yang[2], Xiaochao Fan[1], Ge Ren[1], Yong Yang[1(✉)], and Hongfei Lin[2(✉)]

[1] School of Computer Science and Technology, Xinjiang Normal University, Ürümqi, China
[2] Department of Computer Science and Technology, Dalian University of Technology, Dalian, China
liang@mail.dlut.edu.cn, hflin@dlut.edu.cn

Abstract. People usually believe that using pre-trained word vectors or pre-trained language models can effectively improve task performance. But that is not the case. A sufficient amount of annotated data is usually required to fine-tune the pre-trained language model and pre-trained word vectors for downstream tasks. In addition, the relevance of the training corpus and task corpus also affects task performance to a large extent. In this paper, we systematically compared the effects of different types of pre-trained embeddings and self-training embeddings on the performance of AES. At the same time, we propose an effective solution to the above problem, an automatic essay scoring method that includes pre-trained and self-training word embeddings. We conducted experiments on a public available dataset, including 8 subsets, and the experimental results show the effectiveness of this method.

Keywords: Automatic Essay Scoring · Self-training embeddings · Pre-trained embeddings · Natural language processing

1 Introduction

Automatic Essay Scoring (AES) is a technology that uses linguistics, statistics and natural language processing techniques to automatically score essays, and is often used in large-scale examinations [4]. Since AES was put forward in 1966, it has been successfully applied to various major exams, such as IELTS and TOEFL exams abroad and the College English Test (CET) in China. In addition, AES is an important part of the Automatic Essay Evaluation (AEE) system. Accurately scoring the essay can make the AEE system evaluate the essay more objectively and with reference significance, which can help teachers to better evaluate the students' writing and improve the quality of teaching. AES can effectively avoid the influence of the teacher's subjective factors on the essay scoring, thereby greatly improving the fairness and accuracy of the essay scoring. At the same time, the automatic essay scoring technology can greatly reduce the workload of

© Springer Nature Switzerland AG 2021
H. Lin et al. (Eds.): CCIR 2021, LNCS 13026, pp. 155–167, 2021.
https://doi.org/10.1007/978-3-030-88189-4_12

the marking teacher and save material and manpower costs. In short, AES has a wide range of applications in the fields of examination and education.

Early automatic essay scoring methods constructed shallow features that could reflect the characteristics of the essay [2,25], such as vocabulary, grammar, syntax, and text structure features, and then used machine learning methods for feature mining to indirectly evaluate essay scores. This type of traditional machine learning methods based on feature engineering ignores the latent semantic features of the text and does not truly understand the essay from the semantic level. Therefore, this method cannot achieve satisfactory results. In addition, the traditional machine learning cannot avoid the inherent defects of time-consuming, labor-consuming and weak generalization ability in the feature construction process. Recently, neural network methods based on deep learning have achieved better performance in essay scoring tasks [1,20]. The current deep learning methods are based on the word embedding representation obtained from large-scale corpus training, and use neural networks for feature extraction, crossover and fusion, which can extract the high-dimensional latent semantic features of the essay from a deeper level.

However, most of the corpus used by the commonly used pre-training word vectors comes from online social media, and they are quite different from the essay corpus in terms of semantic expression, logical structure, and language style. In addition, due to the lack of essay corpus, there may not be enough data to fine-tune it. Using these word vectors may not improve the performance of essay scoring, but will cause a certain degree of semantic deviation.

In response to the above problems, we propose an effective solution that contain both pre-trained embeddings and self-training embeddings. And we systematically compare the effects of different commonly used pre-trained word vectors on the performance of AES. In addition, we also investigate the effect of network layer depth on pre-trained word embeddings and self-training word embeddings.

In the following sections, we discuss related work on automatic essay scoring, followed by a description of the methods, experimental data and results of our study.

2   Related Work

Automatic essay scoring is an important auxiliary tool in the field of education and research, and there have been many research results at home and abroad. For the research of AES, according to the different methods of use, this section will sort out the previous work from two aspects: the traditional machine learning method based on feature engineering and the method based on deep learning.

The traditional machine learning method based on feature engineering constructs artificial features according to the grammar and syntax rules of the essay score, and uses traditional machine learning methods such as Logistic Regression (LR) and Support Vector Machine (SVM) to score the essay. Abroad, PEG (Project Essay Grade) [14] is one of the earliest automatic scoring systems, which automatically score essays by constructing the structure of writing and other shallow semantic features. Yannakoudakis [4] uses text length, n-gram features of words and parts of speech as feature sets, and uses SVM as a classifier to score the essay. Pramukantoro et al. [16] proposed an unsupervised automatic essay scoring method based on cosine similarity. Domestically, Liang Maocheng et al. [9] first proposed the AES scoring method, using grammar, syntax and language expression and other essay features, and scoring the essay by using the linear regression method. Zhou Ming et al. [12] extracted essay features from three levels of words, sentences, and paragraphs, used a variety of machine learning algorithms to classify the essay of the text, and used a linear regression model to score the structure of the text. Yu Liqing [10] used a variety of essay features including simple features such as the number of words and sentences, semantic features such as part of speech and n-grams, and constructed an automatic essay scoring system using random forest algorithm. Zhao Ruixue [18] uses the word vector clustering method to obtain the three features of word frequency, vocabulary size and distribution position. At the same time, it uses the fit feature, text feature and non-text feature as the feature set, and finally uses the random forest algorithm to The essay is scored.

In recent years, deep learning methods based on neural networks have achieved many research results in the field of AES. Huang Kai [7] uses convolutional neural networks to obtain sentence beauty features and integrate topic features to score the essay. Taghipour et al. [20] uses a convolutional neural network and a Long Short-Term Memory network in series to automatically extract essay features. Dong et al. [6] used a hierarchical convolutional neural network model based on the attention mechanism to automatically learn features from two levels of sentence structure and text structure and score the essay. The SkipFlow neural network model proposed by Tay et al. [21] can better model the semantic connection of long texts. Liu et al. [11] proposed a Two-Stage Learning Framework (TSLF). First, a deep neural network is used to obtain the semantic representation of the essay. Number, etc.), and then pass it to the XGboost classifier to predict the essay score. Rodriguez et al. [17] applied the BERT (Bidirectional Encoder Representation from Transformers) and XLNet model to the field of automatic essay scoring and achieved good performance. Uto et al. [22] integrated item response theory (IRT) into the deep network model to eliminate the prejudice of the rater when scoring the essay, thereby improving the scoring performance of the model. Li et al. [8] proposed a deep neural network model for cross-topic knowledge transfer, and achieved the best performance in scoring cross-topic essays. Ormerod [13] aims at the problem that large-scale language models are difficult to train, and proposes an efficient language model based on the transformer structure, which can accurately and efficiently predict essay scores. From the perspective of hierarchical semantics, Zhou Xianbing et al. [24] constructed a essay scoring model based on multi-level semantic features.

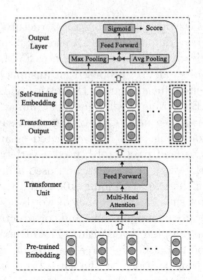

Fig. 1. The overall framework of our proposed method.

3   Method

Our model structure mainly contain a layer of improved Transformer unit and dual pooling operations. At the same time, we take advantage of self-training and pre-trained embeddings to mitigate the semantic deviation caused by insufficient training data. The model structure is shown in Fig. 1. We will make a detailed introduction to each part below.

3.1   Input for Pre-trained Embeddings

This layer maps each word of the input essay to a low-dimensional feature space to obtain a low-dimensional dense vector of the word while maintaining the semantic information of the word. We use the public available pre-trained word vectors and set the embedding representation of the essay $W = \{x_1, x_2, \cdots, x_L\}$, $x_i \in \mathbb{R}^d$ is Word embedding representation, d is the dimension of word vector.

3.2   Transformer Unit

The self-attention mechanism connects any two words in a sentence by calculating the semantic similarity and semantic features of each word in the sentence and other words so as to better obtain the long-distance dependency. The multi-head self-attention proposed by [23] is used in this section. For a given query $Q \in \mathbb{R}^{(n_1 \times d_1)}$, key $K \in \mathbb{R}^{(n_1 \times d_1)}$, value $V \in \mathbb{R}^{(n_1 \times d_1)}$, we use the dot product to calculate attention parameters. The formula is as follows:

$$\text{Attention}(Q, K, V) = \text{softmax}\left(\frac{QK^T}{d_1}\right) V \tag{1}$$

where d_1 is the number of hidden layer unites.

The multi-head attention mechanism maps the input vector X to query, key, and value using linear changes. In our task, key = value. Then, the model learns the semantic features between words through the l-time attention. For the i-th attention head, let the parameter matrix $W_i^Q \in \mathbb{R}^{n_1 \times \frac{d_1}{l}}$, $W_i^K \in \mathbb{R}^{n_1 \times \frac{d_1}{l}}$, $W_i^V \in \mathbb{R}^{n_1 \times \frac{d_1}{l}}$, we use the dot product to calculate the semantic features between them:

$$M_i = \text{Attention}(QW_i^Q, KW_i^K, VW_i^V) \tag{2}$$

The vector representation obtained by the multi-head attention mechanism is concatenated to obtain the final feature representation:

$$H^s = \text{concat}(M_1, M_2, \ldots, M_l) W_o \tag{3}$$

where W_o is the parameter matrix, concat means concatenation operation.

Not like the original Transformer structure, we do not use residual connections in this task. We only use feed forward neural network with ReLU activation function. The calculation formula is as follows:

$$S = Relu(W \cdot H^s + b) \tag{4}$$

where W is the is the parameter matrix, b is the bias term, Relu is the activation function, and S is the semantic representation of the essay.

3.3 Input for Self-training Embeddings

Since the common word embedding representations exhibit a linear structure, that makes it possible to meaningfully combine words by an element-wise addition of their vector representations. In this part, each word is once again initialized randomly as vector $C = (c_1, c_2, \cdots, c_n)$, $c_i \in \mathbb{R}^{d'}$.

In order to better mitigate the semantic deviation caused by pre-trained word vectors, we append the initialized word embeddings to each word representation that is the output of Transformer later. The new representation of each word is $s'_i = s_i \oplus c_i$, where $\oplus$ is the vector concatenation operation.

3.4 Output Module

After the Transformer layer, we use maximum pooling to capture the most significant features of each word embedding dimension, which makes some specific words in the essay have higher value in the word embedding space, and use the average pooling operation to average the embedding space of each word to obtain the overall essay semantics. The double pooling operations can capture different semantic information and complement each other. Therefore, we use fusion technology to fuse three kinds of semantic information to further improve the performance of essay scoring. The calculation formula is as follows:

$$P_{max} = Pooling\_max\left(s'_1, s'_2 \cdots, s'_L\right) \tag{5}$$

$$P_{avg} = Pooling\_average\left(s_1', s_2' \cdots, s_L'\right) \tag{6}$$

$$P_W = Concat\left(P_{max}, P_{avg}\right) \tag{7}$$

where P_{max} and P_{avg} are the feature vectors of two pooling layer, L is the essay length and P_W is the feature vector after the fusion of three kinds of semantic information.

This article uses the feature vector after the two kinds of pooling vectors fusing as the semantic representation of the essay, and then it is passed into a two-layer fully connected network. At the same time, We use ReLU as the activation function of the network to enhance nonlinear representation learning. Finally, we get the score of the essay by using Sigmoid activation function. The calculation process is as follows:

$$P_i = Relu\left(W \cdot P_W + b\right) \tag{8}$$

$$Score = Sigmoid\left(W' \cdot P_i + b'\right) \tag{9}$$

where W and W' is the is the parameter matrix, b and b' is the bias term, Relu and Signoid is the activation function and $Score$ is the final score of the essay.

4   Experiment and Result Analysis

This chapter first introduces the datasets and evaluation method, then explains the experimental setup, compares the performance of the proposed model and the baseline model in detail. At the same time, we also analyze the pre-trained word vector and model efficiency.

4.1   Data and Evaluation Indexes

We use the publicly available datasets of the Kaggle ASAP (Automated Student Assessment Prize) competition, which are widely used in the field of automatic essay scoring. The essays included in ASAP are all written by students in grades 7–10. According to the topic content of the essay, the datasets are divided into 8 subsets. Each subset contains a essay prompt document and multiple related topic essays. The details of the data set are shown in Table 1.

In order to be consistent with the competition and baseline methods, this paper uses Quadratic Weighted Kappa (QWK) as the evaluation index. QWK is a consistency test method used to evaluate whether the results of the model are consistent with the actual results. Assuming that the score of the essay can be divided into N levels, the calculation formula of QWK is as follows:

$$QWK = 1 - \frac{\sum W_{i,j} O_{i,j}}{\sum W_{i,j} E_{i,j}} \tag{10}$$

Table 1. Statistics of the ASAP dataset

Prompt	Essays	Avg length	Scores
D1	1783	350	2–12
D2	1800	350	1–6
D3	1726	150	0–3
D4	1772	150	0–3
D5	1805	150	0–4
D6	1800	150	0–4
D7	1569	250	0–30
D8	723	650	0–60

$$W_{i,j} = \frac{(i-j)^2}{(N-i)^2} \tag{11}$$

where, O is an n-order histogram matrix, $O_{i,j}$ represents the number of essays with an expert score of i and a model score of j, and $W_{i,j}$ represents the difference between the expert score and the model score The second weighted matrix of $E_{i,j}$ represents the product of the probabilities that the expert score is i and the model score is j, in which $E_{i,j}$ and $O_{i,j}$ need to be normalized.

4.2 Experimental Setup

Since the test set of the Kaggle ASAP competition has not been made public, we only use its training datasets as the experimental data for this article. Consistent with the work of [6,20,22], we use a 5-fold cross-validation method in the experiment to evaluate the model we proposed. Each fold has a training data ratio of 60% and a validation set of 20%, The test set accounts for 20%.

In the training process, we use both the pre-trained embeddings and the randomly initialized self-trained embeddings, where the pre-trained word vector comes from Glove Common Crawl (6B token) and the dimension is 300, and the dimension of self-training embeddings is 100. For the Transformer unit, the multi-head attention has 3 heads. The first Feed-Forward network has one layer with 200 neurons. The number of fully connected network nodes in both layers is 100 dimensions. Our optimization function is RMSProp, the decay rate is set to 0.9, and the learning rate is set to 0.001. In order to prevent overfitting, an early stopping mechanism is used during training, and We add dropout with a drop rate of 0.1 in the final layer. In addition, we use different maximum text lengths for different subsets as input to our model. The specific lengths are shown in the Table 2.

4.3 Result Analysis

In order to verify the effectiveness of our method, we compare the following baseline methods:

Table 2. Maximum text length for different subsets

Prompt	Maximum text length
D1	600
D2	600
D3	300
D4	300
D5	300
D6	400
D7	500
D8	800

Table 3. Performance comparison between our method and other baseline methods.

Model	D1	D2	D3	D4	D5	D6	D7	D8	Avg QWK (%)
CNN*	79.70	63.40	64.60	76.70	74.60	75.70	74.60	68.70	72.25
LSTM*	77.50	68.70	68.30	79.50	81.80	81.30	80.50	59.40	74.63
SkipFlow LSTM*	83.20	68.40	69.50	78.80	81.50	81.00	80.00	69.70	76.51
CNN + LSTM*	82.10	68.80	69.40	80.50	80.70	81.90	80.80	64.40	76.08
CNN+LSTM+ATT*	82.20	68.20	67.20	81.40	80.30	81.10	80.10	70.50	76.38
Topic+BiLSTM+ATT*	82.70	69.60	69.10	81.60	81.10	**82.30**	80.90	70.70	77.30
BERT + XLNet*	80.78	69.67	70.31	81.9	80.82	81.45	80.67	60.46	75.78
Electra+Mobile-BERT*	83.10	67.90	69.00	**82.50**	81.70	82.20	**84.10**	**74.80**	78.20
Trm-pre	83.62	71.32	**73.18**	80.68	82.46	81.96	80.58	71.66	78.18
Trm-self	83.92	70.40	73.12	80.98	83.08	81.44	80.12	73.62	78.34
Ours	**84.58**	**72.18**	72.94	81.00	**83.28**	81.62	80.78	72.90	**78.66**

*indicates a direct reference to the original results.

CNN LSTM [20]: Convolutional neural network or long short-term memory network is used to extract essay features and score the essay.

SkipFlow LSTM [21]: The SkipFlow mechanism is added to the LSTM network, which uses the semantic relationship between the hidden layers of the LSTM as an auxiliary feature for essay scoring.

CNN+LSTM [20]: Using the integrated learning method, the prediction results of 10 CNN models and 10 LSTM models are averaged and used as the final prediction result.

CNN+LSTM+ATT [6]: Using CNN+ATT to get the sentence representation of the essay, and use it as the input of LSTM+ATT to get the final semantic representation of the essay.

Topic+BiLSTM+ATT [3]: The two-way long and short-term memory network and the attention mechanism are used to obtain the semantic representation of the essay and the semantic representation of the prompts, and then the vector multiplication is used to obtain the topic relevance of the essay, and finally the topic relevance is integrated into the semantic vector of the essay to score the essay.

BERT+XLNet [17]: Using the integrated learning method, the prediction results of 6 different BERT models and 6 different XLNet models are averaged and used as the final prediction result.

Electra+Mobile-BERT [13]: The high-efficiency language model Electra [5] and Mobile-BERT[19] models are applied to automatic essay scoring, and integrated learning is used to further improve the scoring performance.

Trm-pre: We only use pre-trained embeddings in the first layer.

Trm-self: We only use self-training embeddings in the first layer.

Table 3 lists the comparison between the method we proposed and previous work. The experimental results show that:

(1) The mixed model of CNN and LSTM can effectively improve the overall performance of essay scoring. Compared with a single model, the six subsets have improved. The performance of the model is further improved after using the attention mechanism, which shows that the attention mechanism can effectively obtain the semantics of the essay. The fusion of topic features in the BiLSTM model based on the attention mechanism can effectively improve the performance of essay scoring, indicating that the topic relevance of the essay plays an important role in essay scoring.

(2) Using a large pre-trained language model for automatic essay scoring can achieve certain results. The overall performance of BERT+XLNet is slightly higher than LSTM by 1.15%, indicating that large-scale pre-trained language models are not effective in essays on specific topics. Electra+Mobile-BERT is better than the BERT+XLNet model on 7 subsets, and the overall performance is 2.42% higher, indicating that the lightweight language model is better than the large pre-training language model when performing essay scoring tasks.

(3) Our method achieves the better performance on the datasets than several more complex approaches, including a hybrid of CNN, LSTM and Attention Mechanism and even the recent work Electra+Mobile-BERT. This is probably because of insufficient training samples, which leads to insufficient fine-tuning of the large pre-training model, thus making the model performance poor. Compared with CNN, each subset has been greatly improved and the average performance has increased by 6.41%. In addition, using a combination of pre-trained and self-training embeddings is superior to a single kind of embeddings in the comprehensive performance of the 8 data sets.

4.4 Pre-trained vs Self-training Embeddings

The pre-training word vector is trained from a large-scale corpus and can contain rich semantic information. For most NLP tasks, using word vectors based on large-scale corpus pre-trained can effectively improve model performance. To verify the effect of self-training embeddings and pre-trained embeddings on the AES performance, we conducted a brief experiment on the different pre-trained embeddings and self-training embeddings, which include Glove [15](-glv-6B-300d; -glv-27B-200d; -glv-42B-300d)[1], self-training and self-/pre-train. Where B stands for scale, d stands for dimension, self-training means that randomly initialize the word vectors, and self-/pre-train is the method we used. Except self-/pre-train, all others only contain one embedding layer. We conducted related experiments on the data set D1, and the experimental results are shown in Fig. 2.

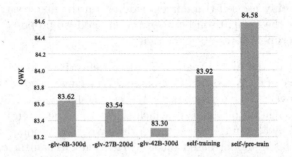

Fig. 2. The AES performance using different embeddings

It can be seen from the figure that when no pre-trained word vector is used, the performance of the model is much higher than using any other pre-trained word vectors. Compared with using the largest-scale pre-trained word vector -glv-42B-300d, the -w/o word-vec scoring performance is still 2.16% higher. This may be because the training corpus of the pre-trained word vector and the essay corpus are quite different in terms of semantic expression, logical structure, and language style. Therefore, using the word vector that pre-trained on this corpus might result in a certain degree of semantic deviation, which reduces the scoring performance of the model. What's more, from the figure, we can also see that when we combine pre-trained and self-training embeddings, AES performance has been greatly improved. The experimental results show that our method is an effective solution to alleviate the semantic bias caused by insufficient data and the weak relevance of the training corpus of the pre-trained word vector.

4.5   Network Layer Analysis

Our intuition is that as the number of network layers deepens, the parameters of the first few layers have less and less influence on the model performance. But that is not the case. We used pre-trained and self-training embeddings as the input of the model to experiment with the transformer structure of 1–6 layers.

---

[1] https://nlp.stanford.edu/projects/glove/.

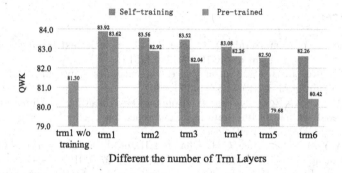

Fig. 3. The relationship between different Trm layers on pre-training and self-training embeddings

At the same time, we also add an additional contrast experiment where the pre-trained word vector parameter is set to untrainable when there is only one layer of Transformer units. The experimental results are shown in Fig. 3.

From the figure, we can see that as the number of model layers deepens, the model performance will gradually decrease. This phenomenon can be understood easily. So when the data is insufficient, we should reduce the depth of the model. In addition, we also found that the gap between self-training and pre-training embeddings is almost getting bigger, which seems to be inconsistent with our understanding. This may be because when the model is deepened, the model cannot fine-tune the word vector well, and at the same time it aggravates the semantic deviation of the pre-trained word vector.

5   Conclusions

We propose an effective method to solve the problem of insufficient data and the large difference in the correlation between the target corpus and the training corpus of the pre-trained word vector. We use pre-trained embeddings as the input of a single-layer transformer unit, and then merge the self-trained embeddings and the output of the transformer unit. In addition, average pooling and maximum pooling are used to obtain full-text semantics from different perspectives. And then fuse these three semantic vectors and pass them into a fully connected neural network using the ReLU activation function for feature interaction. Finally, we use the Sigmoid function to predict the essay score. The experimental results of 8 datasets in the Kaggle ASAP competition show that our method can accurately and efficiently score the essay, and compared with the baseline method, it achieves the best performance on this datasets.

Acknowledgments. This work was supported by grant from the Xinjiang Uygur Autonomous Region Natural Science Foundation Project No. 2021D01B72. This work was also supported by the Natural Science Foundation of China No. 62066044.

References

1. Alikaniotis, D., Yannakoudakis, H., Rei, M.: Automatic text scoring using neural networks. arXiv preprint arXiv:1606.04289 (2016)
2. Chen, H., Xu, J., He, B.: Automated essay scoring by capturing relative writing quality. Comput. J. **57**(9), 1318–1330 (2014)
3. Chen, M., Li, X.: Relevance-based automated essay scoring via hierarchical recurrent model. In: 2018 International Conference on Asian Language Processing (IALP), pp. 378–383. IEEE (2018)
4. Chen, Y.Y., Liu, C.L., Lee, C.H., Chang, T.H., et al.: An unsupervised automated essay-scoring system. IEEE Intell. Syst. **25**(5), 61–67 (2010)
5. Clark, K., Luong, M.T., Le, Q.V., Manning, C.D.: Electra: pre-training text encoders as discriminators rather than generators. arXiv preprint arXiv:2003.10555 (2020)
6. Dong, F., Zhang, Y., Yang, J.: Attention-based recurrent convolutional neural network for automatic essay scoring. In: Proceedings of the 21st Conference on Computational Natural Language Learning (CoNLL 2017), pp. 153–162 (2017)
7. Kai, H.: Research and implementation of key techniques of English automatic essay scoring. Ph.D. thesis, Central China Normal University
8. Li, X., Chen, M., Nie, J.Y.: SEDNN: shared and enhanced deep neural network model for cross-prompt automated essay scoring. Knowl.-Based Syst. **210**, 106491 (2020)
9. Liang, M., Wen, Q.: A critical review and implications of some automated essay scoring systems. Technol. Enhanced Foreign Lang. Educ. (5), 18–24 (2007)
10. Liqing, Y.: Research and implementation of automatic scoring system for English propositional essay. Ph.D. thesis, Central China Normal University (2019)
11. Liu, J., Xu, Y., Zhu, Y.: Automated essay scoring based on two-stage learning. arXiv preprint arXiv:1901.07744 (2019)
12. Ming, Z., Yan-ming, J., Cai-lan, Z., Ning, X.: English automated essay scoring methods based on discourse structure. Comput. Sci. **46**(03), 240–247 (2019)
13. Ormerod, C.M., Malhotra, A., Jafari, A.: Automated essay scoring using efficient transformer-based language models. arXiv preprint arXiv:2102.13136 (2021)
14. Page, E.B.: Grading essays by computer: progress report. In: Proceedings of the Invitational Conference on Testing Problems (1967)
15. Pennington, J., Socher, R., Manning, C.D.: Glove: global vectors for word representation. In: Proceedings of the 2014 Conference on Empirical Methods in Natural Language Processing (EMNLP), pp. 1532–1543 (2014)
16. Pramukantoro, E.S., Fauzi, M.A.: Comparative analysis of string similarity and corpus-based similarity for automatic essay scoring system on e-learning gamification. In: 2016 International Conference on Advanced Computer Science and Information Systems (ICACSIS), pp. 149–155. IEEE (2016)
17. Rodriguez, P.U., Jafari, A., Ormerod, C.M.: Language models and automated essay scoring. arXiv preprint arXiv:1909.09482 (2019)
18. Ruixue, Z.: Study on automatic English composition scoring based on word vector clustering and random forest. Microcomput. Appl. **36**(326(06)), 108–111 (2020)
19. Sun, Z., Yu, H., Song, X., Liu, R., Yang, Y., Zhou, D.: MobileBERT: a compact task-agnostic BERT for resource-limited devices. arXiv preprint arXiv:2004.02984 (2020)
20. Taghipour, K., Ng, H.T.: A neural approach to automated essay scoring. In: Proceedings of the 2016 Conference on Empirical Methods in Natural Language Processing, pp. 1882–1891 (2016)

21. Tay, Y., Phan, M., Tuan, L.A., Hui, S.C.: SkipFlow: incorporating neural coherence features for end-to-end automatic text scoring. In: Proceedings of the AAAI Conference on Artificial Intelligence, vol. 32 (2018)
22. Uto, M., Okano, M.: Robust neural automated essay scoring using item response theory. In: Bittencourt, I.I., Cukurova, M., Muldner, K., Luckin, R., Millán, E. (eds.) AIED 2020. LNCS (LNAI), vol. 12163, pp. 549–561. Springer, Cham (2020). https://doi.org/10.1007/978-3-030-52237-7_44
23. Vaswani, A., et al.: Attention is all you need. arXiv preprint arXiv:1706.03762 (2017)
24. Xianbing1, Z., Xiaochao, F., Ge, R., Yong, Y.: English automated essay scoring methods based on multilevel semantic features. Comput. Appl., 1–8 (2021)
25. Yannakoudakis, H., Briscoe, T., Medlock, B.: A new dataset and method for automatically grading ESOL texts. In: Proceedings of the 49th Annual Meeting of the Association for Computational Linguistics: Human Language Technologies, pp. 180–189 (2011)

Enhanced Hierarchical Structure Features for Automated Essay Scoring

Junteng Ma[1], Xia Li[1,2(✉)], Minping Chen[1], and Weigeng Yang[1]

[1] School of Information Science and Technology, Guangdong University of Foreign Studies, Guangzhou, China
junteng.ma@qq.com, minpingchen@gdufs.edu.cn, 761669265@qq.com
[2] Guangzhou Key Laboratory of Multilingual Intelligent Processing, Guangdong University of Foreign Studies, Guangzhou, China
xiali@gdufs.edu.cn

Abstract. Automated Essay Scoring (AES) aims to evaluate the quality of an essay automatically. In practice, an essay is usually organized in a hierarchical structure, which means that the writer needs to organize the main ideas into different paragraphs, and organize coherent sentences and appropriate words for the essay. Therefore, it is crucial to model the hierarchical structure of essays for AES. For addressing this issue, most of the existing works used neural network-based architectures (e.g., CNNs and LSTMs) to model the hierarchical structure of essays. Different from previous studies, we propose a novel hierarchical graph structure based on graph convolutional networks (GCN) to encode the hierarchical structure of essays and hope to obtain those structured coherence and discourse information from the graph aggregation. We conduct several experiments on ASAP dataset and the experimental results demonstrate the effectiveness of our method.

Keywords: Automated Essay Scoring · Hierarchical structure features · Graph neural network

1 Introduction

Automated Essay Scoring (AES) is a task of automatically evaluating the quality of an essay, which is an important problem in the field of education, especially in language learning and language testing.

One of the core challenges of the AES task is to effectively extract the crucial features related to content, coherence and other metrics of evaluating the quality of an essay. In the previous studies, various solutions are proposed to capture the representative features of essays. Some early studies used carefully designed shallow features, such as essay length, number of sentences and sentence structure features, to express the representation of the essay's quality [11,16,19].

In recent years, many of the neural network-based methods are proposed to obtain the high-level of features related to the quality of the essays. Some works focused on modeling the coherence features. For example, Tay et al. [23] proposed to learn neural coherence features by using a tensor layer which takes each

© Springer Nature Switzerland AG 2021
H. Lin et al. (Eds.): CCIR 2021, LNCS 13026, pp. 168–179, 2021.
https://doi.org/10.1007/978-3-030-88189-4_13

two hidden states of LSTM layer from different time steps as input and calculate the similarity of the input pair. Farag et al. [8] proposed 'a local coherence network to study the coherence feature between sentences by using convolution operation. Some other works focused on modeling the document structure. For example, Dong and Zhang [6] divided the essay into sentences and proposed a hierarchical convolutional neural network, in which the two layers of CNN are used to learn sentence-level representation and document-level representation, respectively. Dong et al. [7] also proposed a hierarchical structure to model the essay, in which CNN is used to extract n-gram features in a sentence and LSTM is used to model the sentences. Zhang and Litman [31] followed the same approach of [7] and used a co-attention mechanism [20] to capture relationships between the essay and source article. Liao et al. [13] considered both coherence feature and hierarchical structure to obtain the high-level coherence representation.

Although the above methods achieved good performance by modeling the coherence or document structure of essays with refined neural network architecture, we find that an essay is typically organized in a hierarchical structure in practice. We argue that explicitly modeling this hierarchical structure is important for AES. To this end, in this paper, we propose a novel hierarchical graph structure based on graph convolutional network to explicitly encode the hierarchical structure of an essay and obtain the structured coherence and discourse information by the aggregation of those different features. We also introduce an interact attention mechanism to conduct an enhancing operation between the aggregated features learned by the graph network and the hierarchical features learned by previous neural models to obtain the enhanced hierarchical structure feature as the final essay representation for the prediction. The contributions of our work are as follows:

(1) We propose a novel hierarchical graph structure based on graph convolutional network to explicitly encode the structure of essays, including nodes of words, sentences, document and different types of edges. To the best of our knowledge, this is the first work to explore modeling document structure with graph convolutional network for automatic essay scoring task.

(2) We also introduce an interact attention mechanism to obtain an enhanced hierarchical structure feature for better representing essay's structure. We conduct several experiments on ASAP dataset and the experimental results demonstrate the effectiveness of our model.

2   Our Method

In this section, we first present an overview of our model in Sect. 2.1 and introduce the two kinds of hierarchical structure features extraction methods in Sect. 2.2 and Sect. 2.3, respectively. Then, we will explain the interact attention which is used to obtain the final enhanced features in Sect. 2.4. Finally, we will give model training of our method in Sect. 2.5.

2.1   Overview Architecture of Our Method

The architecture of our method is shown in Fig. 1. As mentioned before, the purpose of this paper is to explore the contribution of graph neural network for the AES task. To this end, our model consists of three components. The first component is the encoding of essay's document structure using neural network, followed the work of Dong et al. [7], we use a hierarchical attention network to encode and extract the essay's hierarchical features, which is denoted as H_{hier}. The second component is the encoding of essay's document structure using our designed graph neural network, which is denoted as H_G. The third component is an interact attention layer, which is used to aggregate the hierarchical neural features H_{hier} and the graph hierarchical features H_G to obtain a better representation of essay's document structure.

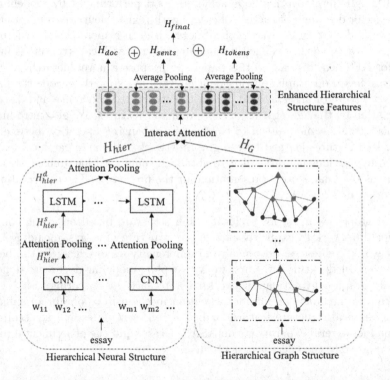

Fig. 1. Overview architecture of our method. We first extract the hierarchical structure features from CNN-LSTM-att network [7] and our designed graph neural network, which are denoted as H_{hier} and H_G. Then we use an interact attention layer to aggregate H_{hier} and H_G, and obtain the enhanced hierarchical structure features H_{doc}, H_{sents} and H_{tokens} for final prediction.

2.2   Hierarchical Neural Feature Representation

Dong et al. [7] proposed an effective hierarchical model called attention-based recurrent convolutional neural network to model the essay. Following the work of [7], we use the CNN-LSTM-att architecture to model the essay and extract three levels of representation to represent essay's document hierarchical structure, which is token-level features H_{hier}^w, sentence-level features H_{hier}^s and document-level feature H_{hier}^d.

In detail, the token-level features are obtained by applying convolution operation on each sentence. Then it is followed by an attention pooling to get the sentence-level features. For document-level feature, we use LSTM and another attention pooling on sentence-level features representation. We concatenate this three types of features as hierarchical neural feature, which is denoted as H_{hier}.

2.3   Hierarchical Graph Feature Representation

In order to obtain the hierarchical structure of essays effectively, we propose a novel graph neural network to encode the different levels of essay's feature. We hope to capture a better representation of essay's document structure by using the explicitly designed graph structure.

Graph Structure. In our model, we represent each essay as a graph, which is denoted as $G = (V, E)$, V is node set and E is edge set. All the graph are represented as undirected graph.

Considering a writer usually organizes an article from top, such as the main idea, to the bottom, such as words, sentences and paragraphs, just like a multi-way tree. To imitate such a structure, we design the graph structure as followed:

- We design three types of heterogeneous nodes to represent an essay's document hierarchical structure, which are document node D, sentence node S and token node T. Note that, since the public ASAP data set has no paragraph information, we don't use paragraph node here.
- We design four types of edges in our proposed graph structure. First, in order to model the hierarchical structure of the essay, we construct edges from document node to its corresponding sentence nodes and edges from sentence node to its corresponding token nodes, which are denoted as (D, S) and (S, T) respectively. Then, considering that the adjacent information can express the local coherent information and the discourse information, we also construct edges from sentence node to its adjacent sentence node and edges from token node to its adjacent token node, which are denoted as (S, S) and (T, T) respectively.

Graph Encoding. Following previous work, we use GCN [10] to encode the graph. GCN is a kind of neural network used on graphs, which aggregates the neighbors of a node in the form of convolution. For a given graph $G = (V, E)$, we first denote the initial node features as $H^0 \in \mathbb{R}^{|V| \times dim}$, where dim is dimension size and $|V|$ is the number of nodes. The node features of l layer of GCN is updated as followed:

$$H^{l+1} = ReLu\left(D^{-\frac{1}{2}}AD^{-\frac{1}{2}}\left(W^{l+1}H^l + B^{l+1}\right)\right) \tag{1}$$

where $W^{l+1} \in \mathbb{R}^{dim_{in} \times dim_{out}}$, $B^{l+1} \in \mathbb{R}^{1 \times dim_{out}}$ are learnable parameters, $A \in \mathbb{R}^{|V| \times |V|}$ is an adjacency matrix that stores edge information. Since the graph we construct is undirected, the adjacency matrix A is symmetric.

$$A(i,j) = \begin{cases} 1, \exists\, (v_i, v_j) \in E, \\ 0, \text{otherwise} \end{cases} \tag{2}$$

The degree matrix $D \in \mathbb{R}^{|V| \times |V|}$ is calculated by A:

$$D(i,j) = \begin{cases} \sum_{j'=1}^{|V|} A\left(j', i\right), i = j \\ 0, \qquad\qquad \text{otherwise} \end{cases} \tag{3}$$

Regarding the initial various types of node features, we use pre-trained word embeddings to perform bottom-up pooling, that is, a sentence node feature is obtained by average pooling of the words node features corresponding to it, and the document node feature is obtained by average pooling of the sentence node features. These initial node features of the graph will be encoded by several layers of GCN, and the output node features of the last layer are denoted as H_G.

2.4 Interact Attention Mechanism

In order to allow the model to focus on different aspects of information, Vaswani et al. [25] project the input hidden states into multiple subspace, which is multi-head attention. In the original Transformer [25], the input of the model is first encoded by a self-attention sublayer as the Eq. (4), where Q, K and V are calculated by three different weight matrices:

$$Attention(Q, K, V) = softmax\left(\frac{QK^T}{\sqrt{dim}}\right) V \tag{4}$$

In this paper, we use an operation similar to self-attention, namely interact-attention, for fusing two types of features more effectively. As described in Sect. 2.2 and Sect. 2.3, given the extracted hierarchical neural feature representation $H_{hier} \in \mathbb{R}^{|V| \times dim}$ and the hierarchical graph feature representation $H_G \in \mathbb{R}^{|V| \times dim}$, where $|V|$ is the number of nodes in the graph, and dim is dimension size. We treat the hierarchical neural feature as Q, and the hierarchical graph feature as K and V, then the interact attention operation between them is calculated as:

$$H_{IA} = InteractAttention(H_{hier}, H_G) = Attention(H_{hier}, H_G, H_G) \tag{5}$$

The output of the model $H_{IA} \in \mathbb{R}^{|V| \times dim}$ contains three types of node features, which are denoted as H_{IA}^{doc}, H_{IA}^{sents} and H_{IA}^{tokens}. We average pooling sentence and token node features separately, which are denoted as H_{sents} and H_{tokens}, and concatenate them with document node features to obtain the final representation H_{final} which is shown in Eq. (6). And it will be activated by *sigmoid* function for predicting essay scores. Note that, following original Transformer's structure [25], we also stack a position–wise feed–forward network (FNN) sublayer after interact-attention layer. In order to prevent gradient vanishing, after this two sublayer, the output will have a residual connections [9] and layer normalization [2] operation.

$$H_{final} = Concat(H_{doc}; H_{sents}; H_{tokens}) \tag{6}$$

2.5 Model Training

We use Mean Square Error (MSE) for model training, which is defined in Eq. (7), where N is the number of essays in a specific prompt, y_i^* is the ground truth score of an essay, and y_i is the predicted score by the model.

$$mse\,(y, y^*) = \frac{1}{N} \sum_{i=1}^{N} (y_i - y_i^*)^2 \tag{7}$$

We train our model on training set and select the best model evaluated on development set according to QWK score, and report the results on test set.

3 Experiments

3.1 Dataset

We use ASAP dataset[1] (Automated Student Assessment Prize) for our experimental evaluation. ASAP dataset consists of 8 different essay prompts, which are originally written by students between Grade 7 and Grade 10. The statistical information is shown in Table 1. ARG, RES, and NAR represent argumentative, response and narrative essays, respectively. We apply 5-fold cross-validation to evaluate models with 60/20/20 split for train, validation, and test sets, which are provided by Taghipour and Ng [22]. Following previous works, we use Quadratic Weighted Kappa (QWK) to evaluate the performance of the model.

3.2 Experiment Settings

We use NLTK to tokenize the data, and convert words into lowercase. When training the model, the scores are scaled into the range [0, 1] and convert back to the original scores at evaluation stage for calculating QWK. As for vocabulary size of the data, we select the most 4,000 frequent words in the training data. We

---

[1] https://www.kaggle.com/c/asap-aes/data.

Table 1. Statistics of the ASAP dataset

Prompt	#Essays	Genre	Avg Len.	Range
1	1783	ARG	350	2–12
2	1800	ARG	350	1–6
3	1726	RES	150	0–3
4	1772	RES	150	0–3
5	1805	RES	150	0–4
6	1800	RES	150	0–4
7	1569	NAR	250	0–30
8	723	NAR	650	0–60

use GloVe [18] word embedding as initial word representation with a dimension of 50. We use RMSprop [4] as the optimizer with initial learning rate 0.001. The model is trained for 30 epochs with batch size 10.

The extraction of hierarchical neural features follows the settings of Dong et al. [7], window size of CNN is 5 with 100 kernels, hidden states of LSTM is 100, and the dropout rate is 0.5. We apply 2 layer GCN operation, and the dimensions of GCN output is 100. We apply 1 layer interact-attention sublayer, the number of heads is 5, the model dimension is 100, and hidden size of FFN is 256.

3.3   Compared Models

Firstly, We use **LSTM** and **BiLSTM** as two basic baselines to be compared. In this two models, we treat the last output and mean pooling of two-layer of LSTM and BiLSTM as the essay representations. Secondly, We use **CNN+LSTM** and **RL1** model proposed by Taghipour and Ng [22] and Wang et al. [26], the former ensemble CNN and LSTM for essay scoring and the latter propose a reinforcement learning based model which using QWK as reward function. Finally, We use three methods that are also used to model essay coherence as compared models. They are **LSTM-CNN-att** proposed by Dong et al. [7], **SKIPFLOW** proposed by Tay et al. [23] and **HierCoh** proposed by Liao et al. [13]. **LSTM-CNN-att** used CNN and LSTM encoding layer and attention mechanism to obtain the final hierarchical essay representation. **SKIPFLOW** integrated the learning of coherence into the model. **HierCoh** proposed a hierarchical coherence model to acquire high-level coherence.

3.4   Experimental Results

Table 2 shows the QWK score of different models on ASAP dataset. Firstly, we compare our model with the first four methods. We can see that our model perform much better on each prompt than the first four baseline methods which only use LSTM/BiLSTM and do not consider any document structure or coherence information.

Table 2. QWK score of different models on ASAP dataset

Methods	Prompt1	Prompt2	Prompt3	Prompt4	Prompt5	Prompt6	Prompt7	Prompt8	Average
LSTM(last)	0.165	0.215	0.231	0.436	0.381	0.299	0.323	0.149	0.275
BiLSTM(last)	0.226	0.276	0.239	0.502	0.375	0.412	0.361	0.188	0.322
LSTM(mean)	0.582	0.517	0.516	0.702	0.604	0.670	0.661	0.566	0.602
BiLSTM(mean)	0.591	0.491	0.498	0.702	0.643	0.692	0.683	0.563	0.608
CNN+LSTM [22]	0.821	0.688	0.694	0.805	0.807	0.819	0.808	0.644	0.761
RL1 [26]	0.766	0.659	0.688	0.778	0.805	0.791	0.760	0.545	0.721
LSTM-CNN-att [7]	0.822	0.682	0.672	0.814	0.803	0.811	0.801	0.705	0.764
SKIPFLOW [23]	0.832	0.684	0.695	0.788	0.815	0.810	0.800	0.697	0.764
HierCoh [13]	0.839	0.702	0.711	0.809	0.801	0.827	0.820	0.631	0.763
Ours	0.834	0.700	0.694	0.820	0.814	0.818	0.809	0.693	**0.773**

Secondly, we compare our model with the two hierarchical models CNN+ LSTM [22] and LSTM-CNN-att [7]. We can see that our method outperforms the CNN+ LSTM [22] and LSTM-CNN-att [7] methods by QWK value of 0.12 and 0.09, respectively, demonstrates the effectiveness of our model. CNN+LSTM [22] model do not use sentence information while we consider the relationship between words and sentences. And our model enhances the QWK on prompt 1–7 to LSTM-CNN-att [7] except prompt 8. It indicates that our designed graph structure and the interact-attention mechanism can make full use of the structure information and achieve better results.

Finally, we compare our model with the coherence based methods SKIPFLOW [23] and HierCoh [13]. The results show that our model performs better than SKIPFLOW [23] on 5 prompts, and surpasses SKIPFLOW [23] and HierCoh [13] by the overall average QWK of 0.09 and 0.1. However, our method performs obviously worse than HierCoh [13] on prompt 3, 6, 7. The essays of this 3 prompts have similar number of average words, may have similar structures. So our model probably cannot leverage the differences of the essays, while HierCoh [13] explicitly model the coherence of the essays.

We note that prompt 8 has the lowest number of essays (723 essays) and the longest average length of essays (650 words) as shown in Table 1, which means that it is more difficult to predict the score for prompt 8. Our model gets slightly lower results than LSTM-CNN-att [7] and SKIPFLOW [23] on prompt 8, but performs more stable than HierCoh [13].

3.5 Visualization Analysis

For further verify the effectiveness of our model, we visualize the final representation of our model. We use t-SNE [14] to project the representation of essays on the same prompt into two dimensions, and draw scatter plots to observe the distribution of the essays.

Figure 2 is comparisons between our model and the model of Dong et al. [7] on prompt 3 and 5. We both use the final representation of two models before fully connected layer for visualization. The score ranges of prompt 3 and 5 are [0–3] and [0–4], respectively. As shown in subfigure (a) and (b), the representations in same score of Dong et al. [7] are more scattered while are gathered much more closer in our model, especially the essays in lower scores. We can see the similar situation in (c) and (d). The visualization proves that our model can make full use of the structure information and obtain better representations for predicting.

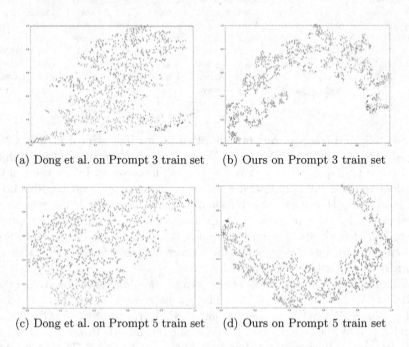

(a) Dong et al. on Prompt 3 train set   (b) Ours on Prompt 3 train set

(c) Dong et al. on Prompt 5 train set   (d) Ours on Prompt 5 train set

Fig. 2. The visualization results of final representations on training set of prompt 3 and prompt 5. (a)(c) and (b)(d) are model of Dong et al. [7] and our model, respectively. Different colors indicate different scores. We can see that the essays with same score are more closer in our model than in CNN-LSTM-att [7], especially the essays in lower score on prompt 3. On prompt 5, it is clear to see that the representations gather more closer in (d) while more scattered in (c).

4   Related Work

Automated essay scoring aims to evaluate the quality of an essay automatically. Early studies mainly focus on the extraction of shallow features of essays to improve the performance, such as essay length, sentence length and syntactic tree features [3,11,16,19].

In recent years, neural network-based methods have achieved promising results in AES task. Some models use CNNs or RNNs to encode the essay from text level [1,22] or hierarchical level [6,7]. For text level, all words of an essay are encoded by CNN or RNN followed by average pooling or attention pooling [21] to generate the essay representation. For hierarchical level, essays are divided into sentences, CNN or RNN are first used to encode the words of each sentence to obtain the sentence representation, then another CNN or RNN takes all sentence representation as input to learn the essay representation. Other models investigate pretrained language models like BERT [5] for AES task. Studies show that directly using pretrained model for AES achieve good performance compared with previous neural models [15,17,24,28].

In addition to encoding essay's content, some studies also integrate the learning of text coherence [8,12,23] and scoring criteria [27,32] into the AES models. Tay et al. [23] propose a tensor layer to learn the semantic relationships between multiple snapshots of LSTM layer, which is regarded as coherence features. Liao et al. [13] propose a hierarchical coherence model to capture the location semantics and high-level coherence. Wang et al. [27] apply reinforcement learning using quadratic weighted kappa as reward function.

In this paper, we also focus on capturing the hierarchical structure features for AES task. Due to good representation of text structure, GCN [10] has been widely used and achieved good results in many NLP tasks, such as text classification [29] and sentiment analysis [30]. Inspired by this advantage, we design a hierarchical graph structure with GCN to encode the hierarchical structure of the essays and obtain structured coherence and discourse information from the graph aggregation. We also introduced a interact attention mechanism to obtain the enhanced hierarchical structure features as the final essay representation.

5   Conclusion

Basing on the observation that the essay is organized in a hierarchical structure, in this paper, we design a novel graph neural network to explicitly encode the hierarchical document structure of essays and introduce an interact attention mechanism to further obtain an enhanced hierarchical features as a better representations for the essays. The experimental results on ASAP data set demonstrate the effectiveness of our method. In the following study, we will focus on the design of different edge types that reflect the quality of essays between heterogeneous nodes in the graph.

Acknowledgements. This work is supported by National Nature Science Foundation of China (61976062) and Special Funds for the Cultivation of Guangdong College Students' Scientific and Technological Innovation (pdjh2021b0177).

References

1. Alikaniotis, D., Yannakoudakis, H., Rei, M.: Automatic text scoring using neural networks. In: Proceedings of the 54th Annual Meeting of the Association for Computational Linguistics (2016)
2. Ba, L.J., Kiros, J.R., Hinton, G.E.: Layer normalization. CoRR abs/1607.06450 (2016). arXiv:1607.06450

3. Chen, H., He, B.: Automated essay scoring by maximizing human-machine agreement. In: Proceedings of the 2013 Conference on Empirical Methods in Natural Language Processing, pp. 1741–1752 (2013)
4. Dauphin, Y.N., de Vries, H., Bengio, Y.: Equilibrated adaptive learning rates for non-convex optimization. In: Proceedings of the 28th International Conference on Neural Information Processing Systems - Volume 1. NIPS 2015, pp. 1504–1512. MIT Press (2015)
5. Devlin, J., Chang, M., Lee, K., Toutanova, K.: BERT: pre-training of deep bidirectional transformers for language understanding. In: Proceedings of NAACL-HLT 2019, pp. 4171–4186. Association for Computational Linguistics (2019)
6. Dong, F., Zhang, Y.: Automatic features for essay scoring - an empirical study. In: Proceedings of EMNLP 2016, pp. 1072–1077 (2016)
7. Dong, F., Zhang, Y., Yang, J.: Attention-based recurrent convolutional neural network for automatic essay scoring. In: Proceedings of CoNLL 2017, pp. 153–162 (2017)
8. Farag, Y., Yannakoudakis, H., Briscoe, T.: Neural automated essay scoring and coherence modeling for adversarially crafted input. In: Proceedings of NAACL-HLT 2018, pp. 263–271 (2018)
9. He, K., Zhang, X., Ren, S., Sun, J.: Deep residual learning for image recognition. In: 2016 IEEE Conference on Computer Vision and Pattern Recognition (CVPR), pp. 770–778 (2016)
10. Kipf, T.N., Welling, M.: Semi-supervised classification with graph convolutional networks. CoRR abs/1609.02907 (2016). arXiv:1609.02907
11. Larkey, L.S.: Automatic essay grading using text categorization techniques. In: Proceedings of the 21st Annual International Conference on Research and Development in Information Retrieval, Melbourne, Australia, 24–28 August 1998, pp. 90–95 (1998)
12. Li, X., Chen, M., Nie, J., Liu, Z., Feng, Z., Cai, Y.: Coherence-based automated essay scoring using self-attention. In: Sun, M., Liu, T., Wang, X., Liu, Z., Liu, Y. (eds.) CCL/NLP-NABD -2018. LNCS (LNAI), vol. 11221, pp. 386–397. Springer, Cham (2018). https://doi.org/10.1007/978-3-030-01716-3_32
13. Liao, D., Xu, J., Li, G., Wang, Y.: Hierarchical coherence modeling for document quality assessment. In: Proceedings of the AAAI 2021, vol. 35, no. 15, pp. 13353–13361, May 2021
14. van der Maaten, L., Hinton, G.: Visualizing data using t-SNE. J. Mach. Learn. Res. 9(86), 2579–2605 (2008)
15. Mayfield, E., Black, A.W.: Should you fine-tune BERT for automated essay scoring? In: Proceedings of the Fifteenth Workshop on Innovative Use of NLP for Building Educational Applications, BEA@ACL 2020, pp. 151–162. Association for Computational Linguistics (2020)
16. McNamara, D., Crossley, S., Roscoe, R., Allen, L.: A hierarchical classification approach to automated essay scoring. Assess. Writ. 23, 35–59 (2015)
17. Ormerod, C.M., Malhotra, A., Jafari, A.: Automated essay scoring using efficient transformer-based language models. CoRR abs/2102.13136 (2021). https://arxiv.org/abs/2102.13136
18. Pennington, J., Socher, R., Manning, C.: GloVe: global vectors for word representation. In: Proceedings of EMNLP 2014, pp. 1532–1543. Association for Computational Linguistics, October 2014
19. Rudner, L.M., Liang, T.: Automated essay scoring using Bayes' theorem. J. Technol. Learn. Assess. 1(2), 1–22 (2002)

20. Seo, M.J., Kembhavi, A., Farhadi, A., Hajishirzi, H.: Bidirectional attention flow for machine comprehension. In: Proceedings of ICLR 2017. OpenReview.net (2017)
21. Sutskever, I., Vinyals, O., Le, Q.V.: Sequence to sequence learning with neural networks. In: Advances in Neural Information Processing Systems 27: Annual Conference on Neural Information Processing Systems 2014, pp. 3104–3112 (2014)
22. Taghipour, K., Ng, H.T.: A neural approach to automated essay scoring. In: Proceedings of EMNLP 2016, pp. 1882–1891 (2016)
23. Tay, Y., Phan, M.C., Tuan, L.A., Hui, S.C.: SkipFlow: incorporating neural coherence features for end-to-end automatic text scoring. In: Proceedings of the Thirty-Second AAAI Conference on Artificial Intelligence, pp. 5948–5955 (2018)
24. Uto, M., Xie, Y., Ueno, M.: Neural automated essay scoring incorporating hand-crafted features. In: Proceedings of the 28th International Conference on Computational Linguistics, pp. 6077–6088. International Committee on Computational Linguistics, December 2020
25. Vaswani, A., et al.: Attention is all you need. In: Proceedings of NIPS 2017. NIPS 2017, pp. 6000–6010. Curran Associates Inc. (2017)
26. Wang, Y., Wei, Z., Zhou, Y., Huang, X.: Automatic essay scoring incorporating rating schema via reinforcement learning. In: Proceedings of the 2018 Conference on Empirical Methods in Natural Language Processing, pp. 791–797. Association for Computational Linguistics, October-November 2018
27. Wang, Y., Wei, Z., Zhou, Y., Huang, X.: Automatic essay scoring incorporating rating schema via reinforcement learning. In: Proceedings of EMNLP 2018, pp. 791–797 (2018)
28. Yang, R., Cao, J., Wen, Z., Wu, Y., He, X.: Enhancing automated essay scoring performance via fine-tuning pre-trained language models with combination of regression and ranking. In: Findings of the Association for Computational Linguistics: EMNLP 2020, pp. 1560–1569. Association for Computational Linguistics, November 2020
29. Yao, L., Mao, C., Luo, Y.: Graph convolutional networks for text classification. In: The Thirty-Third AAAI Conference on Artificial Intelligence. AAAI 2019, pp. 7370–7377. AAAI Press (2019)
30. Zhang, C., Li, Q., Song, D.: Aspect-based sentiment classification with aspect-specific graph convolutional networks. In: Proceedings of EMNLP-IJCNLP 2019, pp. 4567–4577. Association for Computational Linguistics (2019)
31. Zhang, H., Litman, D.J.: Co-attention based neural network for source-dependent essay scoring. In: Proceedings of the Thirteenth Workshop on Innovative Use of NLP for Building Educational Applications@NAACL-HLT 2018, pp. 399–409. Association for Computational Linguistics (2018)
32. Zhao, S., Zhang, Y., Xiong, X., Botelho, A., Heffernan, N.T.: A memory-augmented neural model for automated grading. In: Proceedings of the Fourth ACM Conference on Learning @ Scale, L@S 2017, pp. 189–192 (2017)

IR in Biomedicine

A Drug Repositioning Method Based on Heterogeneous Graph Neural Network

Yu Wang, Shaowu Zhang, Yijia Zhang$^{(\boxtimes)}$, Liang Yang, and Hongfei Lin

School of Computer Science and Technology, Dalian University of Technology, Dalian, China
zhyj@dlut.edu.cn

Abstract. Automated drug repositioning can find potential drugs accurately and reduce R&D costs. To implement a drug repositioning system, first, the author builds a heterogeneous information network of drugs, diseases, and other types of nodes based on the heterogeneous network theory. Second, the meta-path model is introduced, and node and network topology information is learned by deep learning methods. And the interpretability of the model is improved by the attention mechanism. Experimental results on public data sets show that this method has reached state-of-the-art performance, and visual interpretability analysis of one of the inference cases is carried out. At the end of the article, the author provides the potential drugs for Alzheimer's disease inferred from the model and cites relevant literature to prove its effectiveness.

Keywords: Heterogeneous network · Drug repositioning · Graph neural network

1 Introduction

With the development of drug research and development technology, a variety of methods represented by genomics, proteomics, and systems biology have been widely used in the identification of drug targets and the discovery of innovative drugs, but the cycle of research and development is still long and costly. According to statistics, it takes an average of 13.5 years and an investment of 1.8 billion US dollars to develop a new effective drug [1].

At present, the vast majority of drug research and development progress still lags far behind the spread of diseases. For example, the new coronavirus epidemic that broke out at the end of 2019 has now spread to the world. Since effective treatment drugs cannot be developed in a short period, the epidemic has affected the health and economic development of the people in China, South Korea, Japan, Italy, and many other countries, causing huge losses. Therefore, improving the efficiency of drug research and development is related to all mankind.

Drug repositioning refers to the process of mining new indications from existing drugs by using technologies [2]. Drug repositioning can provide effective experimental clues and guidance suggestions for drug research and development, which can enable new drug research and development to break through the limitations of excessive dependence on experimental screening and enter a new stage of combining rational design and

H. Lin et al. (Eds.): CCIR 2021, LNCS 13026, pp. 183–194, 2021.
https://doi.org/10.1007/978-3-030-88189-4_14

experimental verification, thereby greatly shortening the development cycle and reducing the cost of experiments. Drug repositioning has important theoretical value and practical significance for drug development.

For example, Viagra, used to treat erectile dysfunction, launched by Pfizer Pharmaceuticals, whose main ingredient is Sildenafil, was originally developed to treat cardiovascular diseases such as angina pectoris and hypertension. However, it was unexpectedly discovered in clinical trials that Sildenafil has a good therapeutic effect on erectile dysfunction.

2 Related Works

Early drug repositioning cases were mostly due to accidental discoveries, and later research scholars proposed a series of automated methods based on machine learning and network analysis. Gottlieb et al. [3] used features to calculate the similarity between drugs and the similarities between diseases to predict potentially treatable diseases of drugs. Zeng et al. [4] used an autoencoder to fuse the features of multiple similarity networks and used a variational autoencoder to predict the fused features to implement drug repositioning. Liu et al. [5] used the heterogeneous graph convolutional network method to predict the potential drugs of COVID-19. The above methods have their shortcomings. For example, feature-based methods are time-consuming and labor-intensive, and some feature engineering requires medical-related domain knowledge; machine learning-based methods lack interpretability and cannot provide persuasive evidence for drug developers.

In the entity network, there are a large number of types and patterns of association paths between drugs and diseases. The role of learning and distinguishing different types of association paths in establishing the relationship between drugs and disease is the key to improving the ability of drug repositioning.

Drug repositioning not only needs to efficiently and accurately predict the implicit interaction between drugs and diseases with high credibility in a complex entity relationship network but also requires the interpretability for the mechanism of interaction between drugs and diseases.

Currently, heterogeneous graph neural networks (GNN) are widely used in recommendation systems. Fan et al. [6] designed a meta path-guided heterogeneous GNN to learn the embeddings of objects in Taobao's recommendation service. Luo et al. [7] propose a multi-layer GNN model which can take historical actions and the surrounding environment into account to do prediction. Heterogeneous network theory can reasonably represent and distinguish different types of entities and entity relationships in the network, and improve the ability of inference.

This article will construct a heterogeneous network for drug repositioning based on heterogeneous theory, and integrate multiple types of entities; Introduce meta-path model and attention mechanism to improve the inference ability and Interpretability of drug repositioning model based on deep learning methods. Specifically, it is divided into two stages: first, establish a drug and disease prediction model based on the meta-path model and multi-head attention mechanism to predict the implicit relationship between the drug and the disease; then, based on the prediction results, with meta-path instances and attention weights, establish an explanation path for the mechanism of interaction of drugs and diseases.

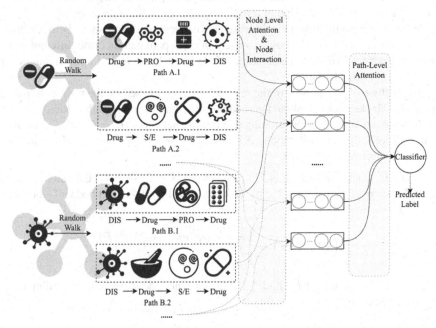

Fig. 1. Drug repositioning frame diagram based on heterogeneous graph neural network

3   Methodology

The overall architecture of the model proposed in this paper can be seen in Fig. 1. Based on the deep learning method, this paper introduces the meta-path model and attention mechanism. First, the paths are sampled by random walk algorithm according to the predefined meta-path pattern in the constructed heterogeneous network. Second, the semantic information of the path and the embedding of the nodes can be learned by node interactions, thereby the role of different types and patterns of association paths can be effectively learned and distinguished. The attention mechanism is divided into node-level attention and path-level attention, which can capture important node interaction information and important path information, and improve the interpretability of the drug repositioning model.

3.1   Meta-path Model

The meta-path model was proposed by the team of Professor Jiawei Han of the University of Illinois in VLDB2011 [8]. Its advantage is that it can effectively distinguish and integrate different types of path information in heterogeneous networks.

Meta-path refers to the path pattern in the drug repositioning relationship network, which is abstracted from specific associated path instances and is represented in the form of $N_1 \xrightarrow{E_1} N_2 \xrightarrow{E_2} \ldots \xrightarrow{E_{n-1}} N_n$, where $N_1 \sim N_n$ denote the entity node type, and $E_1 \sim E_{n-1}$ denote the edge, that is, the type of relationship that exists between entities.

In the drug repositioning entity-relationship network, the relationship between drugs and diseases is established through a variety of meta-path patterns, and different meta-path patterns have different meanings and effects in establishing the implicit relationship between drugs and diseases.

The paths will be sampled in the relational network based on the meta-path model randomly. The key is to combine the random walk with the meta-path model.

Assume that the drug repositioning relationship network is $G = (N, E, T)$, where N denotes a collection of entity nodes, E denotes a collection of entity relationships, and T is a collection of entity types.

When the meta path mode is $\mathcal{P} = N_1 \xrightarrow{E_1} N_2 \xrightarrow{E_2} \ldots \xrightarrow{E_{n-1}} N_n$, the sampling path reaches node n^i at the i-th step, the transition probability to the next node n^{i+1} is shown in Eq. (1).

$$p\left(n^{i+1}|n^i, \mathcal{P}\right) = \begin{cases} \frac{1}{|N(n^i, \mathcal{P})|}, & if n^{i+1} \in N\left(n^i, \mathcal{P}\right) \\ 0, & else \end{cases} \tag{1}$$

$N(n^i, \mathcal{P})$ denotes the neighbor nodes that the current node n^i can visit according to the meta-path mode $\mathcal{P}$.

3.2 Node Interaction Based on Convolution

Most methods based on heterogeneous networks use network representation learning to find key nodes and meta-paths and capture complex structures. To further mine the information between nodes, this paper adopts the method of node interaction proposed by Jin et al. [9].

The paths are sampled by random walk algorithm, and node interaction between the paths is performed to obtain the semantic information, which mainly includes three behaviors, namely Shift, Hadamard Product, and Sum. The process is similar to Convolutional Neural Network (CNN). The difference is that the weight of the convolution kernel of the CNN is a trainable parameter, however, there is no trainable convolution kernel in the convolution operation of node interaction.

Figure 2 is a schematic diagram of calculating the node interaction between the two paths. The paths are generated by random walks starting from the target drug and the target disease respectively.

First, we need to reverse one of the paths, and then move it from left to right. Take the nodes of the overlapping parts of the two paths to calculate the Hadamard product and sum them.

Then, repeat the operations of moving, Hadamard product, and summing until there are no overlapping nodes between the paths.

The calculation process of a single node interaction can be expressed by Eq. (2):

$$E_n = \sum_{i=0}^{n-1} e_{1,i} \circ e_{2,L_2-i-1} \tag{2}$$

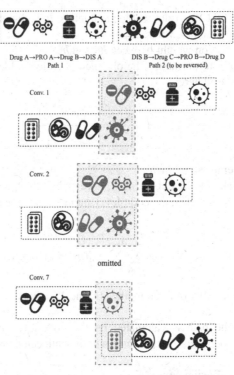

Fig. 2. Schematic diagram of node interaction based on convolution

where L_2 denotes the length of the second path, $e_{1,i}$ denotes the embedding of the i-th node in the first path and ∘ denotes the Hadamard product. Then the path embedding H can be obtained by concatenating the embedding E processed by node interaction:

$$H = \left[E_0 \oplus E_1 \oplus \cdots \oplus E_{L1+L2-2} \right] \tag{3}$$

where L denotes the length of the element path, $\oplus$ denotes the concatenating operation of the vector, $H \in \mathbb{R}^{L \times E}$, due to the moving operation in the convolution, $L = L_1 + L_2 - 1$, E is the dimension of the node embedding vector.

3.3 Attention Mechanism

To enhance the inference ability and interpretability of the model, this paper introduces the attention mechanism, which specifically includes a path-level attention fusion module and a node-level attention module.

Node-Level Attention Module

The goal of the node-level attention mechanism is to learn the weights of different types of nodes in the path, as shown by Eq. (4):

$$h_{0j} = (h_0 W_T)^T \cdot (h_j W_S) \tag{4}$$

Among them, W_s and W_t are trainable parameters, h_i is the element obtained after the i-th convolution interaction, where $j = 0, 1, \ldots, L - 1$.

h_{0j} denotes the attention weight between h_0 and h_j.

Then it is necessary to normalize the obtained attention value, and the calculation method of softmax is used here:

$$\alpha_j = \text{softmax}\left(\frac{h_{0j}}{\iota}\right) = \frac{\exp\left(\frac{h_{0j}}{\iota}\right)}{\sum_{i=0}^{L-1} \exp\left(\frac{h_{0j}}{\iota}\right)} \tag{5}$$

where ι denotes the temperature factor.

In addition, to make the model learning more stable, this paper uses a multi-head attention mechanism to calculate the node-level attention, and the calculation process is shown by Eq. (6):

$$z = \sigma\left(W_q \cdot \frac{1}{K} \sum_{n=1}^{K} \sum_{j=0}^{L-1} \alpha_{jn}(h_j W_{Cn}) + b_q\right) \tag{6}$$

where K denotes the number of attention heads, and W_q, W_{Cn}, and b_q are all trainable parameters.

Therefore, after the path is calculated by the node-level attention module, the semantic information between different nodes in the path can be better captured.

Path-Level Attention Fusion Module

The path-level attention mechanism will calculate the attention weight of the path instance based on the features of the path instance.

First, the non-linear transformation is performed on the embedding obtained by the node-level attention module.

The attention weight of each path can be considered as the importance of the path, and the calculation process is shown by Eq. (7):

$$\omega = w^T \cdot \tanh\left(W_p \cdot z + b_p\right) \tag{7}$$

where w^T, W_p, b_p denote trainable parameters respectively, and then similar to the node-level attention mechanism, all attention weights are normalized through softmax just like Eq. (5).

The attention weight of a path can explain the importance of each path in a certain inference process. The attention weight and the semantic embedding of the path are fused and calculated. The final aggregate embedding Z of all paths can be obtained by Eq. (8):

$$Z = \sum_{j=0}^{P-1} \beta_j \cdot z_j \tag{8}$$

where β_j denotes the attention weight of path j.

3.4 Loss Function

The final prediction result can be obtained through a linear transformation and a softmax classifier with aggregate embedding Z, and the loss can be calculated according to the prediction result and the true label. The loss function is shown in Eq. (9):

$$L\left(Y, \widehat{Y}\right) = -\sum_i \left(y_i \log \widehat{y_i} + (1 - y_i)\log\left(1 - \widehat{y_i}\right)\right) \tag{9}$$

where y_i denotes the true label of sample i, and $\widehat{y_i}$ denotes the predicted label probability of the sample.

4 Experiment and Analysis

In this section, we mainly show the details of the experimental setup and the results of the comparative experiment. To illustrate the effectiveness and interpretability of our method, we conduct a visualization and interpretability analysis on an inference example, and in addition, take Alzheimer's disease as an example to make inferences about potential drugs.

4.1 Data Set

The experimental part of this paper uses the open-source data contributed by Zeng et al. [6], and the data comes from two widely used databases: DrugBank [10] and repoDB [11]. This data set includes four relational networks and several similarity networks calculated based on artificially defined rules. This experiment only uses the binary relation networks, and whose description is shown in Table 1.

Table 1. Data set description and statistics

Net	Shape	Num	Sparsity	Description
Drug - disease	1519 × 1229	6,677	0.003577	Therapeutic relationship
Drug - drug	1519 × 1519	290,836	0.126047	Clinically reported DDI
Drug - protein	1519 × 1025	6,744	0.004331	Target protein of drugs
Drug - S/E	1519 × 12904	382,041	0.019491	Side effects of drugs

4.2 Experimental Setup

This paper uses five-fold cross-validation to evaluate the performance of the method. The positive cases are all drug-disease pairs that have a therapeutic relationship, and the negative cases are obtained by random sampling, and the number is equal to the positive cases. In the experiment, five meta-path patterns are manually pre-defined, and the specific parameters are shown in Table 2.

In order to illustrate the effectiveness of the methods used in this article, the following methods are selected as the baseline model for comparison:

Table 2. Parameter setting

Parameter	Value
Batch size	128
Temperature factor	0.2
Learning rate	0.001
Head number	3
Path sample number	13
Meta-path pattern 1	drug → disease
Meta-path pattern 2	drug → disease → drug → disease
Meta-path pattern 3	drug → protein → drug → disease
Meta-path pattern 4	drug → drug → drug → disease
Meta-path pattern 5	drug → S/E → drug → disease

- DTINet [12]: Learn the features of nodes in a heterogeneous network, and use inductive matrix completion for prediction.
- KBMF [13]: A method based on kernelized Bayesian matrix factorization, which can take advantage of side information.
- SVM [14]: Use support vector machine for classification.
- deepDR [6]: The autoencoder is used to integrate the relationship network and the artificially defined auxiliary information similarity network feature, and the variational autoencoder is used to predict the relationship.
- Graph CNN: The model is a graph neural network method that uses a trainable convolution kernel, for comparing with the method described in Sect. 2.2. The effectiveness of the node interaction method is proved by ablation experiments.

The experimental results are shown in Table 3. Compared with the previous best method (deepDR), the Graph CNN method using the meta-path model and attention mechanism can obtain higher inference ability without introducing artificially defined side information networks. In addition, the node interaction method based on convolution can further improve the inference ability of the model to reach state-of-the-art performance.

4.3 Interpretability Analysis

The path-level attention fusion module and node-level attention module introduced in this paper also endow the model with interpretability, so as to provide a more persuasive suggestion for artificial decision-making.

To explore the interpretability of the model, we visualized the attention weight of the model in the process of one inference instance, as shown in Fig. 3, where the darker the color, the higher the weight.

It can be seen from the path-level attention weight that the model infers that the relationship between Cefazolin (DB01327) and Staphylococcus aureus infection

Table 3. Model performance comparison

Method	AUROC	AUPR	Acc
DTINet	0.862	0.892	–
KBMF	0.791	0.826	–
SVM	0.771	0.778	–
deepDR	0.908	0.923	0.672
Graph CNN	0.938	0.930	0.866
Graph Inter	**0.965**	**0.966**	**0.918**

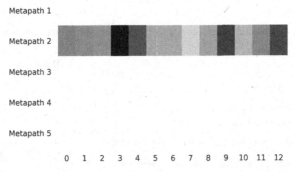

Fig. 3. Path-level attention visualization

(C1318973) is mainly based on meta-path mode 2, namely drug → disease → Drug → disease. By examining the data set, it is found that there are many common therapeutic drugs between this disease and other diseases. The model infers that cefazolin also has a therapeutic effect on Staphylococcus aureus infection based on the multi-hop paths.

Attention graph of node interaction can be observed to further explore the reason why the model makes such an inference. Figure 4 shows the attention weight of the node interaction in the above example. It can be seen that the node interaction behavior of index 3 gets a larger attention weight.

After reviewing the data, it is found that the drug and disease nodes contained in the path have more common neighbor nodes, and the target drug and the target disease can establish a new connection with fewer jumps. By consulting the relevant literature, it is found that there are indeed relevant experiments that prove the effectiveness of the drug [15, 16].

4.4 Example of Drug Repositioning

In order to explore the actual value of the model more objectively, we conducted an inference experiment on new drug discovery for Alzheimer's disease. And the most probable drug candidates are listed in Table 4.

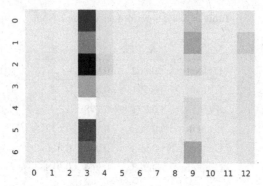

Fig. 4. Node interactive attention map

Table 4. Potential drugs for Alzheimer's disease inferred from the model

Candidate	Rank	Ref
DB00476: Duloxetine	1	[17]
DB01109: Heparin	2	[18]
DB00569: Fondaparinux sodium	3	[19]
DB09525: "Sodium phosphate, monobasic, monohydrate"	4	–
DB00581: Lactulose	5	[20]
DB03793: Cholic Acid	6	[21, 22]

According to relevant public experimental reports or papers, it can be found that many drugs have undergone clinical trials and have been proved to be effective. It can be seen that the drug examples obtained by the model have a certain reference value.

5 Conclusion

This paper describes a method based on a heterogeneous graph neural network to predict the potential therapeutic relationship between drugs and diseases. The method in this paper has reached the state-of-the-art performance on the public data set, and an interpretability analysis has been carried out on one of the inference cases.

In our future work, we will use text mining to mine the relationship between nodes from biomedical texts to expand the structural information of the heterogeneous network and improve the accuracy of inference.

Acknowledgements. This work is supported by grant from the Natural Science Foundation of China (No. 62072070).

References

1. Paul, S.M., et al.: How to improve R&D productivity: the pharmaceutical industry's grand challenge. Nat. Rev. Drug Discov. **9**, 203–214 (2010)
2. Li, J., Zheng, S., Chen, B., Butte, A.J., Swamidass, S.J., Lu, Z.: A survey of current trends in computational drug repositioning. Brief. Bioinform. **17**, 2–12 (2016)
3. Gottlieb, A., Stein, G.Y., Ruppin, E., Sharan, R.: PREDICT: a method for inferring novel drug indications with application to personalized medicine. Mol. Syst. Biol. **7**, 496 (2011). https://doi.org/10.1038/msb.2011.26
4. Zeng, X., Zhu, S., Liu, X., Zhou, Y., Nussinov, R., Cheng, F.: deepDR: a network-based deep learning approach to in silico drug repositioning. Bioinform. **35**, 5191–5198 (2019). https://doi.org/10.1093/bioinformatics/btz418
5. Liu, H., et al.: Drug repositioning for SARS-CoV-2 based on graph neural network. In: 2020 IEEE International Conference on Bioinformatics and Biomedicine (BIBM), pp. 319–322 (2020). https://doi.org/10.1109/BIBM49941.2020.9313236
6. Fan, S., et al.: Metapath-guided heterogeneous graph neural network for intent recommendation. In: Proceedings of the 25th ACM SIGKDD International Conference on Knowledge Discovery & Data Mining, pp. 2478–2486 (2019)
7. Luo, W., et al.: Dynamic heterogeneous graph neural network for real-time event prediction. In: Proceedings of the 26th ACM SIGKDD International Conference on Knowledge Discovery & Data Mining, pp. 3213–3223 (2020)
8. Sun, Y., Han, J., Yan, X., Yu, P.S., Wu, T.: Pathsim: meta path-based top-k similarity search in heterogeneous information networks. Proc. VLDB Endow. **4**, 992–1003 (2011)
9. Jin, J., et al.: An efficient neighborhood-based interaction model for recommendation on heterogeneous graph. In: Proceedings of the 26th ACM SIGKDD International Conference on Knowledge Discovery & Data Mining, pp. 75–84 (2020)
10. Wishart, D.S., et al.: DrugBank 5.0: a major update to the DrugBank database for 2018. Nucleic Acids Res. **46**, D1074–D1082 (2018)
11. Brown, A.S., Patel, C.J.: A standard database for drug repositioning. Sci. Data **4**, 1–7 (2017)
12. Luo, Y., et al.: A network integration approach for drug-target interaction prediction and computational drug repositioning from heterogeneous information. Nat. Commun. **8**, 1–13 (2017)
13. Gönen, M., Khan, S., Kaski, S.: Kernelized Bayesian matrix factorization. In: International Conference on Machine Learning, pp. 864–872. PMLR (2013)
14. Cortes, C., Vapnik, V.: Support-vector networks. Mach. Learn. **20**, 273–297 (1995)
15. Heinrich-Heine University, Duesseldorf: Early Oral Switch Therapy in Low-risk Staphylococcus Aureus Bloodstream Infection. Clinicaltrials.gov (2020)
16. Weis, S., et al.: Cefazolin versus anti-staphylococcal penicillins for the treatment of patients with Staphylococcus aureus bacteraemia. Clin. Microbiol. Infect. **25**, 818–827 (2019)
17. Raskin, J., et al.: Efficacy of duloxetine on cognition, depression, and pain in elderly patients with major depressive disorder: an 8-week, double-blind placebo-controlled trial. AJP **164**, 900–909 (2007). https://doi.org/10.1176/ajp.2007.164.6.900
18. Bergamaschini, L., Rossi, E., Vergani, C., De Simoni, M.G.: Alzheimer's disease: another target for heparin therapy. Sci. World J. **9**, 891–908 (2009)
19. Grossmann, K.: Anticoagulants for treatment of Alzheimer's disease. J Alzheimers Dis. **77**, 1373–1382 (2020). https://doi.org/10.3233/JAD-200610
20. Lee, Y.-S., et al.: Prebiotic lactulose ameliorates the cognitive deficit in Alzheimer's disease mouse model through macroautophagy and chaperone-mediated autophagy pathways. J. Agric. Food Chem. **69**, 2422–2437 (2021). https://doi.org/10.1021/acs.jafc.0c07327

21. Rahman, M.A., Hossain, M.S., Abdullah, N., Aminudin, N.: Validation of Ganoderma lucidum against hypercholesterolemia and Alzheimer's disease. Eur. J. Biol. Res. **10**, 314–325 (2020)
22. Anwar, S., et al.: Structural and biochemical investigation of MARK4 inhibitory potential of cholic acid: towards therapeutic implications in neurodegenerative diseases. Int. J. Biol. Macromol. **161**, 596–604 (2020)

Auto-learning Convolution-Based Graph Convolutional Network for Medical Relation Extraction

Mengyuan Qian, Jian Wang[(✉)], Hongfei Lin, Di Zhao, Yijia Zhang,
Wentai Tang, and Zhihao Yang

School of Computer Science and Technology, Dalian University of Technology,
Dalian 116024, Liaoning, People's Republic of China
{dutqianmengyuan,zhao_di,tangwentai}@mail.dlut.edu.cn,
{wangjian,hflin,zhyj,yangzh}@dlut.edu.cn

Abstract. Medical relation extraction discovers relations between entity mentions in unstructured text, such as biomedical literature. Dependency structures have proven to be useful for this task. However, how to effectively make use of structural information from dependency forests remains a challenging research question. Existing approaches directly employing weighted graphs or variable graphs, where the graph can be viewed as a dependency forest, may not always yield optimal results. In this work, we propose a novel model, the auto-learning convolution-based graph convolutional network (AC-GCN), which learns weighted graphs using a 2D convolutional network. The convolution operation is performed over dependency forests to obtain highly informative features useful for medical relation extraction. Results obtained on three biomedical benchmarks show that our model is able to better learn the structural information of the dependency forests, providing significantly better results than those of previous approaches.

Keywords: Convolutional neural network · Medical relation extraction · Graph convolutional network.

1 Introduction

Medical relation extraction is a task that detects relations among entities that are associated with biological processes from natural language medical texts. The research literature has a wealth of relevant knowledge, and it is growing at an astonishing rate. Therefore, we must accelerate automated methods to improve their relation extraction performances. Additionally, the development of relation extraction tools enables many downstream applications, ranging from question answering to automated knowledge base construction. In the biomedical domain, such tools can help doctors make accurate decisions by mining supportive or contradictory evidence from biomedical research literature [1,2]. Consider the following example: the relations between the drug, gene and mutation. *"The*

ⓒ Springer Nature Switzerland AG 2021
H. Lin et al. (Eds.): CCIR 2021, LNCS 13026, pp. 195–207, 2021.
https://doi.org/10.1007/978-3-030-88189-4_15

*deletion mutation on exon 19 of the **EGFR** gene was present in 16 patients, while the **L858E** point mutation on exon 21 was noted in 10. All patients were treated with **gefitinib** and showed a partial response."*. The two sentences convey the fact that there is a ternary interaction between the the three bold entities. Namely, tumors with the L858E mutation in the EGFR gene can be treated with gefitinib.

With the rise of neural networks, deep learning-based models have become prevalent methods for relation extraction. Most existing models can be categorized into two classes: sequence-based and dependency-based models. Recent advances using sequence-based models via distributional word representations have yielded improvements in relation extraction. Compared to sequence-based models, graph-based models have been shown to be effective in learning long-distance dependencies present in text [3]. Dependency structures have been proven beneficial for relation extraction, as they are able to represent nonlocal syntactic relations. Along this line of thought, in a standard binary relation, the document feature is generally defined in terms of the shortest dependency path between the two entities. The shortest dependency path is able to effectively make use of relevant information while ignoring irrelevant information [4]. However, generalizing this process to an n-ary setting is challenging, as there are $\frac{n(n-1)}{2}$ paths. To overcome this challenge, Peng et al. [2] and Song et al. [5] captured cross-sentence n-ary relation mentions by representing texts with a dependency tree that consisted of both intra- and cross-sentence links between words. With this graphical representation, they applied graph neural networks to encode dependency trees in the medical domain. Although state-of-the-art results are obtained using dependency syntax methods, they require external tools to build a graph for the text. Moreover, the dependency parsing accuracy of this type of approach is relatively low in the medical domain [6]. This can lead to severe error propagation in downstream relation extraction tasks. To alleviate this problem, Jin et al. [7] leveraged full dependency forests for relation extraction, where a full dependency forest is used to encode all possible trees. This method also merges a parser into a relation extraction model so that the parser can be jointly updated based on end-task loss. Guo et al. [8] treated the dependency structure as a latent variable and induced it from unstructured text in an end-to-end fashion.

Dependency structures have proven to be useful for medical relation extraction. However, the weighted graphs generated directly by dependency forests cannot capture the information well. So we further mine rich local and nonlocal dependency information from each full weight graph using a 2D convolutional neural network (CNN). These weights can be viewed as the strengths of the relatedness values between nodes, which can be learned in an end-to-end fashion by using a structured attention mechanism [9]. The convolution operator has been shown to be effective in handling dense multidimensional data, such as images and videos, has exactly these properties: it is parameter efficient and fast to compute [10,11]. Our main idea is to let a CNN learn automatically from a forest through end-to-end training. Next, graph neural networks are used for encoding the forest, which in turn provides features for relation extraction. To the best of our knowledge, this is the first work to use a 2D CNN to learn multidimensional weight graph representations for relation extraction.

Experiments show that our model is able to achieve better performances on various medical relation extraction tasks than those of other approaches. The results obtained on the BioCreative VI ChemProt (CPR) [12] and phenotype-gene relations (PGR) [13] datasets show that our method outperforms existing state-of-the-art methods that use matrix trees and decreases the required computational complexity. For cross-sentence n-ary relation extraction, our model also achieves comparable performance relative to the AGGCN and LF-GCN models.

2   Model

In this section, we formally describe the architecture of the AC-GCN model. As shown in Fig. 1, our AC-GCN framework includes five parts. First, a Bi-LSTM captures the context information. Then, adjacency matrices are constructed by using a multi-head attention mechanism. Next, a 2D convolution automatically learns a multidimensional weight graph. Afterthat, a separate GCN is employed for each graph to encode dependency information. Finally, the outputs of all GCNs are concatenated.

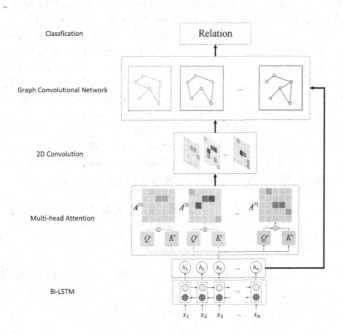

Fig. 1. Model architecture of the proposed AC-GCN.

2.1   Bi-LSTM

Given a document s, each word w_i in it is fed to the context encoder, which outputs the contextualized representations. We first transform the document into a sequence of embeddings $x_1, x_2, ..., x_n$ using a pretrained word embedding procedure. Next, we employ a Bi-LSTM encoder to capture the context information in the vector sequence, and this is then further used to represent the current node of the graph:

$$\overleftarrow{h}_i^{(0)} = \mathbf{LSTM}_l \left(\overleftarrow{h}_{i+1}^{(0)}, x_i \right) \tag{1}$$

$$\overrightarrow{h}_i^{(0)} = \mathbf{LSTM}_r \left(\overrightarrow{h}_{i-1}^{(0)}, x_i \right) \tag{2}$$

where $\overleftarrow{h}_i^{(0)}$, $\overleftarrow{h}_{i+1}^{(0)}$, $\overrightarrow{h}_i^{(0)}$ and $\overrightarrow{h}_{i-1}^{(0)}$ represent the hidden representations of the i-th, $(i\text{-}1)$-th and $(i+1)$-th tokens in document s in two directions.

The state of each word w_i is generated by concatenating the states in both directions:

$$h_i^{(0)} = \left[\overleftarrow{h}_i^{(0)}; \overrightarrow{h}_i^{(0)} \right] \tag{3}$$

2.2   Multi-head Attention

A prebuilt dependency tree requires sophisticated tools that have been trained on manual annotations. Although such methods have demonstrated promising results in relation extraction, they are computationally expensive and not applicable to low-resource languages. As a next step, our model treats the dependency tree as a variable weight graph and constructs it in an end-to-end fashion. We construct each weight graph by using a multi-head attention mechanism [14], which allows the model to integrate different subspace feature representations. Therefore, no additional computational overhead is involved. A multi-weight graph is fed into the 2D CNN to obtain N weight graphs to capture the central dependencies in different representation subspaces. We define k parallel heads to learn the vector values of different channels. Given the input h as a query $Q \in \mathbb{R}^{n \times d}$ and key $K \in \mathbb{R}^{n \times d}$, we compute a weight graph as:

$$A^{(k)} = \mathrm{softmax} \left(\frac{QW_k^Q \times \left(KW_k^K \right)^T}{\sqrt{d}} \right) \tag{4}$$

where $W_k^Q \in \mathbb{R}^{d \times d}$ and $W_k^K \in \mathbb{R}^{d \times d}$ are trainable parameters of the k-th head. $\sqrt{d}$ is the scaling factor. A_{ij} denotes the normalized attention score between the i-th token and the j-th token with h_i and h_j, respectively.

2.3   2D-CNN

As the multi-head attention mechanism generates N weight graph representations, we treat the multi-weight graphs $A^{(1)}, A^{(2)}, ..., A^{(N)}$ as a whole feature

$A^* \in \mathbb{R}^{N \times l \times l}$, where l is the length of the sentence and N is the number of channels in the convolution block. The main characteristic of our model is that the dependency information is defined by a convolution over 2D shaped weight graphs. We choose to apply convolutional operations to the multi-weight graphs to extract useful features from this structure. The model uses A^* as an input for a 2D convolutional layer with filters ω. In the next step, the kernels are applied to each convolutional region, and such a layer returns a feature map tensor is:

$$A_c^* = \text{Sigmoid}\,(A^* * \omega + b_c) \tag{5}$$

where b_c is a trainable parameter representing the kernel bias. We use padding to keep the feature lengths of the outputs the same as those of the inputs. To further encode different subspace representations, the refined multidimensional weight graphs are transformed into N different fully connected weighted graphs. The resulting matrices are fed into N separate GCN layers, generating new node representations.

2.4 GCN

Graph neural networks (GCNs) have been widely used in various natural language processing tasks to encode context [15], as they are able to model long-term dependency relations based on an information aggregation scheme. In this paper, we also employ a GCN to encode the weight graphs. The GCN block takes node embeddings and an adjacency matrix that represents the graph as inputs. The hidden node embedding obtained from the Bi-LSTM layer and the graphical structure obtained in the convolution step are provided as inputs for the GCN layer. Specifically, the node information is calculated by a graph convolution as follows:

$$h_{k_i}^l = \sigma \left(\sum_{j=1}^{n} A_{ij}^k W_k^l h_i^{l-1} + b_k^l \right) \tag{6}$$

where W_k^l and b_k^l are the weight matrix and bias vector for the k-th weight graph in the l-th layer, respectively. σ is the ReLU activation function.

2.5 Relation Prediction

Next, following Xu et al. [16], we employ a layer aggregation, in which all the subspace outputs are concatenated and fed into a feed-forward layer:

$$h_{com} = W_{com} h_{all} + b_{com} \tag{7}$$

where h_{all} denotes the combination of all subspaces. $W_{com} \in \mathbb{R}^{d \times N \times d}$ is a weight matrix and b_{com} is the bias.

The goal of relation extraction is to predict relations among entities. The information close to the entity tokens in the weight graph is often central to relation classification. Therefore, we concatenate the sentence representation and entity representations to obtain the final representation for classification purposes. We obtain the sentence representation vector directly by:

$$h_{sent} = \text{maxpool}\,[h_{com}] \tag{8}$$

where the max pooling function maps the output vectors to the sentence vector. Then, the i-th entity representation is given by:

$$e_i = \text{maxpool}\,[h_{com}^i] \tag{9}$$

Finally, we integrate all the features by concatenating the final representations of the sentence and entities as follows:

$$h_{final} = [h_{sent}; e_1; \ldots; e_i] \tag{10}$$

The final feature vector is used for classification and is fed to a fully connected softmax layer to obtain a probability distribution over the relation labels. Our model uses cross-entropy loss during training. Our competitive advantage lies in that we achieve better performance than those of other methods without extra parameters and complex structures.

3   Experiments

3.1   Data

We evaluate our AC-GCN model with three datasets on two tasks, namely, cross-sentence n-ary relation extraction and sentence-level relation extraction.

For the cross-sentence n-ary relation extraction task, we use the dataset introduced by Peng et al. [2], which contains 6,987 ternary relation instances and 6,087 binary relation instances extracted from PubMed. Most instances contain multiple sentences, and each instance is assigned one of the five possible labels, including "resistance or nonresponse", "sensitivity", "response", "resistance" and "None". The n-ary and binary relation extraction tasks are divided into cross-sentence and single-sentence relation extraction. For binary-class relations, we follow Peng et al. [2] and Song et al. [5], who grouped all of the relation classes as positive instances and treated "None" as negative.

For the sentence-level relation extraction task, the performance of our model is evaluated on two biomedical datasets. The CHEMPROT corpus consists of PubMed abstracts manually annotated with chemical compound mentions, gene/protein mentions and chemical compound-protein relations [12]. It contains 16,107 training instances, 10,030 development instances and 14,269 testing instances.

The Phenotype-Gene Relations (PGR) corpus, a silver standard corpus of human phenotypes, gene annotations and their relations [13], is also used. It contains 11,780 training instances and 219 test instances, and we separate the first 15% of the training instances as our development set.

3.2 Setup

For the cross-sentence n-ary relation extraction task, the test accuracies averaged over five cross-validation folds are reported in accordance with previous evaluation methods. For the sentence-level relation extraction task, we report the F1 scores for CPR and PGR. The hyperparameters of the model are tuned using the validation set. We use a 300-dimensional GloVe vector to initialize the word embeddings. These embeddings are fixed during relation extraction training. To prevent overfitting, we use dropout in the GCN and LSTM layers with a dropout rate equal to 0.5. Adam is used as the optimizer. The batch size is set as 50 for all experiments. For each run, we retain the model that achieves the highest F1 score or accuracy on the development set and evaluate and report its score on the test set.

3.3 Main Results

To study the effectiveness of the proposed model, we compare our method with the following baselines:

- Feature-Based: An LSTM model based on the shortest dependency paths between all entity pairs [1].
- Tree LSTM: A tree LSTM model that combines dependency tree information with other lexical information [17].
- DAG LSTM: Peng et al. [2] used the dependency graph constructed by connecting the roots of dependency trees corresponding to the input sentences.
- Att-GRU: Liu et al. [18] incorporated attention layers at the top of the GRU and in the sequence embedding layers.
- Bran: Verga et al. [19] adopted multi-instance learning based on a biaffine self-attention model to extract relations.
- BioBERT: Lee et al. [20] used bidirectional encoder representations from a Transformers model pretrained on large-scale biomedical corpora.
- GRN: Song et al. [5] encoded a graph by using graph recurrent networks.
- GCN: Zhang et al. [3] encoded a graph of trees pruned by using graph convolutional networks.
- AGGCN: Guo et al. [21] treated attention matrices as the adjacency matrices of forests.
- Variant GRN: Song et al. [22] used multi-forests to replace with best tree structure.
- DDCNN: Jin et al. [7] built a forest by a learnable dependency parser.
- LF-GCN: Guo et al. [8] induced the dependency structure automatically with a variant of the matrix tree theorem.

3.4 Cross-Sentence n-Ary Relation Extraction

For the cross-sentence n-ary relation extraction task, Table 1 shows the performance comparison between the AC-GCN model and all baselines. All neural network-based graph structures outperform the feature-based classifier, illustrating their advantage in handling sparse linguistics without requiring intense feature engineering. Graph neural networks based on LSTM (Tree LSTM, DAG LSTM and GRN) continuously improve the relation extraction performances of models. Compared with LSTM-based GNNs, graph convolutional networks based on convolutions are more suitable for relation extraction. Both types of neural networks confirm the usefulness of dependency structures and the effectiveness of GNNs or GCNs in encoding these structures. Models with pruning strategies tend to achieve better results than those without such strategies. Examining the full tree structure, Zhang et al. [3] adopted a rule-based pruning method that produces a strong result. Furthermore, Guo et al. [8,21] provided two soft pruning methods for yielding optimal performances. This suggests that relation extraction may not require full dependency tree features. For ternary relation extraction (first two columns in Table 1), our AC-GCN model achieves accuracies of 88.8 and 88.8 on instances within a single sentence (Single) and on all instances (Cross), respectively, and it outperforms all the baselines. Compared to the state-of-the-art LF-GCN models, our model obtains 3.1- and 1.7-point improvements in terms of multiclass relation extraction, showing that the use of a 2D convolution considerably helps perform fine-grained classification. Although the result of the AC-GCN is 0.6 points worse than that of the LF-GCN for binary cross-sentence relation extraction, our model significantly improves the efficiency of the weight graph refining process. The details of this finding will be discussed in the efficiency section.

Table 1. Average test accuracies according to five-fold validation for binary-class n-ary relation extraction and multi-class n-ary relation extraction. "Ternary" and "Binary" denote ternary drug-gene-mutation interactions and binary drug-mutation interactions, respectively. "Single" and "Cross" indicate accuracies reported on instances within single sentences and considering all instances, respectively.

Model	Binary-class				Multi-class	
	Ternary		Binary		Ternary	Binary
	Single	Cross	Single	Cross	Cross	Cross
Feature-Based [1]	74.7	77.7	73.9	75.2	–	–
Tree LSTM [17]	–	–	75.9	75.9	–	–
DAG LSTM [2]	77.9	80.7	74.3	76.5	–	–
GRN [5]	80.3	83.2	83.5	83.6	71.7	71.7
GCN (Full tree) [3]	84.3	84.8	84.2	83.6	77.5	74.3
GCN (Pruned tree) [3]	85.8	85.8	83.8	83.7	78.1	73.6
AGGCN [21]	87.1	87.0	85.2	85.6	79.7	77.4
LF-GCN [8]	88.0	88.4	86.7	**87.1**	81.5	79.3
AC-GCN (Ours)	**88.8**	**88.8**	**86.8**	86.5	**84.6**	**81.0**

3.5   Sentence-Level Relation Extraction

The results on the CPR [12] and PGR [13] datasets are reported for the sentence-level relation extraction task.

Table 2 shows the main comparison results on the CPR test set, with comparisons between the previous state-of-the-art methods and our model. Compared to the sequence-based model (first two rows in Table 2), the models based on full dependency trees as inputs are able to significantly improve the relation extraction results (from third to fifth rows in Table 2). Models based on dependency forests yield better performances than those using dependency trees (last eight rows in Table 2), and this confirms that the use of forests information can obtain more informative latent structure for relation extraction. The performance of our AC-GCN model is 1.6 points higher than that of the state-of-the-art LF-GCN in terms of their F1 scores, verifying that the use of a 2D convolution function for computing edge weights is highly appropriate for the GCN framework. In addition, we want to understand the contribution of the 2D convolution employed in the model. We conduct an ablation test by removing the 2D convolution component. The ablated model is Att-GCN; the performance of Att-GCN is 1.6 points lower than that of the AC-GCN in terms of their F1 scores, demonstrating the strong contribution of the 2D convolution.

Table 3 shows a comparison between our model and those of previous works on the PGR test set, where our model achieves an F1 score of 92.4, which is better than those yielded by the existing models. We can find that performances of GCN based methods are much better than BioBERT and the result suggest that modeling structure in the input sentence is beneficial to the relation extraction task. Leveraging dependency forests (last seven rows in Table 3), such models significantly outperform those using dependency trees. This further confirms the usefulness of dependency forests for medical relation extraction. Compared to the ablated model (Att-GCN) and the state-of-the-art LF-GCN, our AC-GCN model obtains 1 and 0.5 point score improvements, respectively. These results suggest that the 2D convolution is able to capture task-specific information for improved relation extraction. Compared to the DDCNN, which directly mines useful knowledge from each forest using a conventional CNN, 2D CNNs can extract more useful features from multidimensional forests.

3.6   Analysis

Performance Versus Sentence Length
Figure 2 shows the test accuracies obtained with different sentence lengths. We split the CPR test set into three categories ((0, 50], (50, 100], (100-) based on sentence length. We can see that Att-GCN and the AC-GCN show performance decreases as the input sentence lengths increase. This is likely because longer sentences correspond to more complex dependency structures. The AC-GCN is consistently better than Att-GCN, and the gap is larger on longer instances. This demonstrates that a 2D convolution is highly effective in modeling complex nonlocal interactions for improved prediction.

Table 2. Test results on the CPR dataset. The result labeled with "*" is obtained based on the authors' released model, as we do not have published results for the dataset.

Model	F1
Att-GRU [18]	49.5
Bran [19]	50.8
GCN [3]	52.2
Tree-DDCNN [7]	50.3
Tree-GRN [7]	51.4
Edgewise-GRN [22]	53.4
KBest-GRN [22]	52.4
AGGCN [21]	56.7
DDCNN [7]	55.7
LF-GCN [8]	58.9
Att-GCN [14]	64.2
LF-GCN* [8]	64.2
AC-GCN (Ours)	**65.8**

Table 3. Test results on the PGR dataset.

Model	F1
BioBERT [20]	67.2
GCN [3]	81.3
Tree-GRN [7]	78.9
Edgewise-GRN [22]	83.6
KBest-GRN [22]	85.7
AGGCN [21]	89.3
DDCNN [7]	89.3
Att-GCN [14]	91.4
LF-GCN [8]	91.9
AC-GCN (Ours)	**92.4**

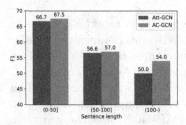

Fig. 2. F1 score versus sentence length.

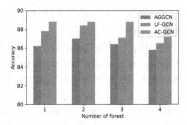

Fig. 3. Accuracy versus the number of forests on Peng's dataset. The results of the AGGCN and LF-GCN are reproduced based on their released codes.

Performance Versus Number of Forests

Figure 3 shows the performances of the LF-GCN, AGGCN and AC-GCN with different numbers of forests. Intuitively, it is more beneficial to model graphs with different subspaces because these subspaces can allow for more efficient information integration. The performances of both the AGGCN and LF-GCN increase as the number of forests increases (the optimal number is 2), and this finding coincides with the above intuition. However, the AC-GCN exhibits more advantages than both of the other models with a minimal or maximal number of forests, further demonstrating the superiority of the 2D convolution in utilizing dependency forest.

Efficiency

Table 4 shows the training and decoding times of both the baseline and our model. Our model is 4 to 8 times faster than the baseline in terms of training and decoding speeds, respectively. This is because the convolution operator has the advantage of fast computation speed when processing high-density multi-dimensional data. Our AC-GCN model significantly improves the efficiency of the inference process. On the other hand, this model also yields a performance increase over the state-of-the-art LF-GCN models. Considering the accuracy and efficiency of the AC-GCN model, we expect it to be very effective in practice.

Table 4. The average time of training and decoding iterations in one epoch on CPR dataset. The speed is measured on a single NVIDIA TITAN Xp GPU with a batch size of 50.

Model	Train	Dev
LF-GCN [8]	102 s	33 s
AC-GCN (Ours)	25 s	4 s

Case Study

The proposed approach constructs the weight graph for a given text using an attention mechanism. Therefore, we try to visualize the graph by plotting a heat map of the weight graph obtained by the model. Figure 4 depicts the two heat maps for the sentence "The IL-18 production is located upstream of the cytokine cascade activated by simvastatin.". The sentence expresses the CPR:3 relation between the subject **simvastatin** and the object **cytokine**. From Fig. 4(a), we observe that the AC-GCN can infer connections from the target entities to most of the other words, while Att-GCN in Fig. 4(b) cannot. Furthermore, these two heat maps show different weights between tokens, confirming that a 2D convolution can learn highly useful dependency information.

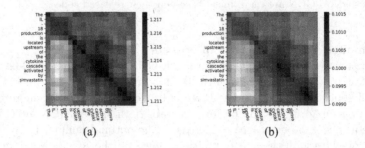

(a)   (b)

Fig. 4. Heat maps of the two weight graphs.

4   Conclusion

In this paper, we present a simple and effective approach for end-to-end medical relation extraction. Our model auto-learns the graph representation for a given text using a 2D convolution. Our experiments show that it considerably outperforms previous state-of-the-art methods on three standard benchmarks. We show through detailed analysis that the advantage of our model comes from the 2D convolution, which can be effectively applied for obtaining weight graph representations.

Acknowledgement. This work was financially supported by the National Natural Science Foundation of China (No. 62072070).

References

1. Quirk C, Poon H.: Distant supervision for relation extraction beyond the sentence boundary. In: EACL, pp. 1171–1182 (2017)
2. Peng, N., Poon, H., Quirk, C., Toutanova, K., Yih, W.T.: Cross sentence N-ary relation extraction with graph LSTMs. Trans. Assoc. Comput. Linguist. **5**, 101–115 (2017)
3. Zhang, Y., Qi, P., Manning, C. D.: Graph convolution over pruned dependency trees improves relation extraction. In: EMNLP, pp. 2205–2215 (2018)

4. Xu, Y., Mou, L., Li, G., Chen, Y., Peng, H., Jin, Z.: Classifying relations via long short term memory networks along shortest dependency paths. In: EMNLP, pp. 1785–1794 (2015)
5. Song, L., Zhang, Y., Wang, Z., Gildea, D.: N-ary relation extraction using graph-state LSTM. In: EMNLP, pp. 2226–2235 (2018)
6. Lease, M., Charniak, E.: Parsing biomedical literature. In: Dale, R., Wong, K.-F., Su, J., Kwong, O.Y. (eds.) IJCNLP 2005. LNCS (LNAI), vol. 3651, pp. 58–69. Springer, Heidelberg (2005). https://doi.org/10.1007/11562214_6
7. Jin, L., Song, L., Zhang, Y., Xu, K., Ma, W.Y., Yu, D.: Relation extraction exploiting full dependency forests. In: AAAI, vol. 34, no. 05, pp. 8034–8041 (2020)
8. Guo, Z., Nan, G., Lu, W., Cohen, S.B.: Learning latent forests for medical relation extraction. In: IJCAI, pp. 3651–3657 (2020)
9. Kim, Y., Denton, C., Hoang, L., Rush, A.M.: Structured attention networks. In: ICLR (2017)
10. LeCun, Y., Bengio, Y.: Convolutional networks for images, speech, and time series. In: The Handbook of Brain Theory and Neural Networks, vol. 3361, no. 10, p. 1995 (1995)
11. Krizhevsky, A., Sutskever, I., Hinton, G.E.: Imagenet classification with deep convolutional neural networks. Commun. ACM 60(6), 84–90 (2017)
12. Krallinger, M., et al.: Overview of the biocreative VI chemical-protein interaction track. In: Proceedings of the Sixth BioCreative Challenge Evaluation Workshop, vol. 1, pp. 141–146 (2017)
13. Sousa, D., Lamurias, A., Couto, F.M.: A silver standard corpus of human phenotype-gene relations. In: NAACL-HLT, pp. 1487–1492 (2019)
14. Vaswani, A., et al.: Attention is all you need. In: Advances in Neural Information Processing Systems, pp. 5998–6008 (2017)
15. Kipf, T.N., Welling, M.: Semi-supervised classification with graph convolutional networks. In: ICLR (2017)
16. Xu, K., Li, C., Tian, Y., Sonobe, T., Kawarabayashi, K.I., Jegelka, S.: Representation learning on graphs with jumping knowledge networks. In: ICML, pp. 5453–5462 (2018)
17. Miwa, M., Bansal, M.: End-to-end relation extraction using LSTMs on sequences and tree structures. In: ACL, pp. 1105–1116 (2016)
18. Liu, S., et al.: Extracting chemical-protein relations using attention-based neural networks. Database J. Biol. Databases Curation 2018, bay102 (2018)
19. Verga, P., Strubell, E., McCallum, A.: Simultaneously self-attending to all mentions for full-abstract biological relation extraction. In: NAACL-HLT, pp. 872–884 (2018)
20. Lee, J., et al.: BioBERT: a pre-trained biomedical language representation model for biomedical text mining. Bioinformatics 36(4), 1234–1240 (2020)
21. Guo, Z., Zhang, Y., Lu, W.: Attention guided graph convolutional networks for relation extraction. In: ACL, pp. 241–251 (2019)
22. Song, L., Zhang, Y., Gildea, D., Yu, M., Wang, Z., Su, J.: Leveraging dependency forest for neural medical relation extraction. In: EMNLP-IJCNLP, pp. 208–218 (2019)

Author Index

Ai, Zhou   59
An, Bo   112

Chen, Chong   141
Chen, Fei   74
Chen, Minping   168
Chen, Zhigang   16

Ding, Kun   129
Dou, Zhicheng   3

Fan, Xiaochao   155

Hao, Wei   59
Hou, Lei   88
Hou, Shuo   74
Hu, Guoping   16

Li, Juanzi   88
Li, Xia   168
Lin, Hongfei   155, 183, 195
Lin, Yuan   129
Liu, Yiqun   45
Liu, Yu   100
Liu, Zhaohui   141
Lu, Mingyu   74

Ma, Junteng   168
Ma, Shaoping   45
Ma, Weizhi   45
Mingyu, Lu   59

Peng, Jingyu   32

Qian, Mengyuan   195
Qin, Xubo   3

Ren, Ge   155

Shi, Jiaxin   88
Shi, Kaijie   88

Shiu, Yingwai   45
Su, Qi   141
Sun, Peijie   32

Tang, Wentai   195

Wang, Baoxin   16
Wang, Beichen   16
Wang, Jian   195
Wang, Kaiqiao   129
Wang, Kaixin   100
Wang, Meng   32
Wang, Shijin   16
Wang, Yu   183
Wang, Ziyue   16
Wen, Ji-Rong   3
Wu, Dayong   16
Wu, Le   32

Xu, Xiujuan   100

Yan, Hongfei   141
Yang, Liang   129, 155, 183
Yang, Weigeng   168
Yang, Yong   155
Yang, Zhihao   195
Yijia, Zhang   59

Zhang, Fuyao   74
Zhang, Min   45
Zhang, Shaowu   183
Zhang, Yijia   74, 183, 195
Zhao, Di   195
Zhao, Xinhang   129
Zhao, Zhehuan   100
Zhou, Xianbing   155
Zhou, Yu   88
Zhu, Yutao   3

Printed in the United States
by Baker & Taylor Publisher Services